W9-BIZ-997

HOLT SCIENCE & TECHNOLOGY

Environmental Science

HOLT, RINEHART AND WINSTON

A Harcourt Classroom Education Company

Austin • New York • Orlando • Atlanta • San Francisco • Boston • Dallas • Toronto • London

Acknowledgments

Chapter Writers

Katy Z. Allen
Science Writer and Former Biology Teacher
Wayland, Massachusetts

Linda Ruth Berg, Ph.D.
Adjunct Professor–Natural Sciences
St. Petersburg Junior College
St. Petersburg, Florida

Jennie Dusheck
Science Writer
Santa Cruz, California

Mark F. Taylor, Ph.D.
Associate Professor of Biology
Baylor University
Waco, Texas

Lab Writers

Diana Scheidle Bartos
Science Consultant and Educator
Diana Scheidle Bartos, L.L.C.
Lakewood, Colorado

Carl Benson
General Science Teacher
Plains High School
Plains, Montana

Charlotte Blassingame
Technology Coordinator
White Station Middle School
Memphis, Tennessee

Marsha Carver
Science Teacher and Dept. Chair
McLean County High School
Calhoun, Kentucky

Kenneth E. Creese
Science Teacher
White Mountain Junior High School
Rock Springs, Wyoming

Linda Culp
Science Teacher and Dept. Chair
Thorndale High School
Thorndale, Texas

James Deaver
Science Teacher and Dept. Chair
West Point High School
West Point, Nebraska

Frank McKinney, Ph.D.
Professor of Geology
Appalachian State University
Boone, North Carolina

Alyson Mike
Science Teacher
East Valley Middle School
East Helena, Montana

C. Ford Morishita
Biology Teacher
Clackamas High School
Milwaukie, Oregon

Patricia D. Morrell, Ph.D.
Assistant Professor, School of Education
University of Portland
Portland, Oregon

Hilary C. Olson, Ph.D.
Research Associate
Institute for Geophysics
The University of Texas
Austin, Texas

James B. Pulley
Science Editor and Former Science Teacher
Liberty High School
Liberty, Missouri

Denice Lee Sandefur
Science Chairperson
Nucla High School
Nucla, Colorado

Patti Soderberg
Science Writer
The BioQUEST Curriculum Consortium
Beloit College
Beloit, Wisconsin

Phillip Vavala
Science Teacher and Dept. Chair
Salesianum School
Wilmington, Delaware

Albert C. Wartski
Biology Teacher
Chapel Hill High School
Chapel Hill, North Carolina

Lynn Marie Wartski
Science Writer and Former Science Teacher
Hillsborough, North Carolina

Ivora D. Washington
Science Teacher and Dept. Chair
Hyattsville Middle School
Washington, D.C.

Academic Reviewers

Renato J. Aguilera, Ph.D.
Associate Professor
Department of Molecular, Cell, and Developmental Biology
University of California
Los Angeles, California

David M. Armstrong, Ph.D.
Professor of Biology
Department of E.P.O. Biology
University of Colorado
Boulder, Colorado

Alissa Arp, Ph.D.
Director and Professor of Environmental Studies
Romberg Tiburon Center
San Francisco State University
Tiburon, California

Russell M. Brengelman
Professor of Physics
Morehead State University
Morehead, Kentucky

John A. Brockhaus, Ph.D.
Director of Mapping, Charting, and Geodesy Program
Department of Geography and Environmental Engineering
United States Military Academy
West Point, New York

Linda K. Butler, Ph.D.
Lecturer of Biological Sciences
The University of Texas
Austin, Texas

Barry Chernoff, Ph.D.
Associate Curator
Division of Fishes
The Field Museum of Natural History
Chicago, Illinois

Donna Greenwood Crenshaw, Ph.D.
Instructor
Department of Biology
Duke University
Durham, North Carolina

Hugh Crenshaw, Ph.D.
Assistant Professor of Zoology
Duke University
Durham, North Carolina

Joe W. Crim, Ph.D.
Professor of Biology
University of Georgia
Athens, Georgia

Peter Demmin, Ed.D.
Former Science Teacher and Chair
Amherst Central High School
Amherst, New York

Joseph L. Graves, Jr., Ph.D.
Associate Professor of Evolutionary Biology
Arizona State University West
Phoenix, Arizona

William B. Guggino, Ph.D.
Professor of Physiology and Pediatrics
The Johns Hopkins University School of Medicine
Baltimore, Maryland

David Haig, Ph.D.
Assistant Professor of Biology
Department of Organismic and Evolutionary Biology
Harvard University
Cambridge, Massachusetts

Roy W. Hann, Jr., Ph.D.
Professor of Civil Engineering
Texas A&M University
College Station, Texas

Printed in the United States of America

ISBN 0-03-064782-7

1 2 3 4 5 6 7 048 05 04 03 02 01 00

Acknowledgments (cont.)

John E. Hoover, Ph.D.
Associate Professor of Biology
Millersville University
Millersville, Pennsylvania

Joan E. N. Hudson, Ph.D.
Associate Professor of Biological Sciences
Sam Houston State University
Huntsville, Texas

Laurie Jackson-Grusby, Ph.D.
Research Scientist and Doctoral Associate
Whitehead Institute for Biomedical Research
Massachusetts Institute of Technology
Cambridge, Massachusetts

George M. Langford, Ph.D.
Professor of Biological Sciences
Dartmouth College
Hanover, New Hampshire

Melanie C. Lewis, Ph.D.
Professor of Biology, Retired
Southwest Texas State University
San Marcos, Texas

V. Patteson Lombardi, Ph.D.
Research Assistant Professor of Biology
Department of Biology
University of Oregon
Eugene, Oregon

Glen Longley, Ph.D.
Professor of Biology and Director of the Edwards Aquifer Research Center
Southwest Texas State University
San Marcos, Texas

William F. McComas, Ph.D.
Director of the Center to Advance Science Education
University of Southern California
Los Angeles, California

LaMoine L. Motz, Ph.D.
Coordinator of Science Education
Oakland County Schools
Waterford, Michigan

Nancy Parker, Ph.D.
Associate Professor of Biology
Southern Illinois University
Edwardsville, Illinois

Barron S. Rector, Ph.D.
Associate Professor and Extension Range Specialist
Texas Agricultural Extension Service
Texas A&M University
College Station, Texas

Peter Sheridan, Ph.D.
Professor of Chemistry
Colgate University
Hamilton, New York

Miles R. Silman, Ph.D.
Assistant Professor of Biology
Wake Forest University
Winston-Salem, North Carolina

Neil Simister, Ph.D.
Associate Professor of Biology
Department of Life Sciences
Brandeis University
Waltham, Massachusetts

Lee Smith, Ph.D.
Curriculum Writer
MDL Information Systems, Inc.
San Leandro, California

Robert G. Steen, Ph.D.
Manager, Rat Genome Project
Whitehead Institute—Center for Genome Research
Massachusetts Institute of Technology
Cambridge, Massachusetts

Martin VanDyke, Ph.D.
Professor of Chemistry, Emeritus
Front Range Community College
Westminister, Colorado

E. Peter Volpe, Ph.D.
Professor of Medical Genetics
Mercer University School of Medicine
Macon, Georgia

Harold K. Voris, Ph.D.
Curator and Head
Division of Amphibians and Reptiles
The Field Museum of Natural History
Chicago, Illinois

Mollie Walton
Biology Instructor
El Paso Community College
El Paso, Texas

Peter Wetherwax, Ph.D.
Professor of Biology
University of Oregon
Eugene, Oregon

Mary K. Wicksten, Ph.D.
Professor of Biology
Texas A&M University
College Station, Texas

R. Stimson Wilcox, Ph.D.
Associate Professor of Biology
Department of Biological Sciences
Binghamton University
Binghamton, New York

Conrad M. Zapanta, Ph.D.
Research Engineer
Sulzer Carbomedics, Inc.
Austin, Texas

Safety Reviewer

Jack Gerlovich, Ph.D.
Associate Professor
School of Education
Drake University
Des Moines, Iowa

Teacher Reviewers

Barry L. Bishop
Science Teacher and Dept. Chair
San Rafael Junior High School
Ferron, Utah

Carol A. Bornhorst
Science Teacher and Dept. Chair
Bonita Vista Middle School
Chula Vista, California

Paul Boyle
Science Teacher
Perry Heights Middle School
Evansville, Indiana

Yvonne Brannum
Science Teacher and Dept. Chair
Hine Junior High School
Washington, D.C.

Gladys Cherniak
Science Teacher
St. Paul's Episcopal School
Mobile, Alabama

James Chin
Science Teacher
Frank A. Day Middle School
Newtonville, Massachusetts

Kenneth Creese
Science Teacher
White Mountain Junior High School
Rock Springs, Wyoming

Linda A. Culp
Science Teacher and Dept. Chair
Thorndale High School
Thorndale, Texas

Georgiann Delgadillo
Science Teacher
East Valley Continuous Curriculum School
Spokane, Washington

Alonda Droege
Biology Teacher
Evergreen High School
Seattle, Washington

Michael J. DuPré
Curriculum Specialist
Rush Henrietta Junior-Senior High School
Henrietta, New York

Rebecca Ferguson
Science Teacher
North Ridge Middle School
North Richland Hills, Texas

Susan Gorman
Science Teacher
North Ridge Middle School
North Richland Hills, Texas

Gary Habeeb
Science Mentor
Sierra-Plumas Joint Unified School District
Downieville, California

Karma Houston-Hughes
Science Mentor
Kyrene Middle School
Tempe, Arizona

Roberta Jacobowitz
Science Teacher
C. W. Otto Middle School
Lansing, Michigan

Kerry A. Johnson
Science Teacher
Isbell Middle School
Santa Paula, California

M. R. Penny Kisiah
Science Teacher and Dept. Chair
Fairview Middle School
Tallahassee, Florida

Kathy LaRoe
Science Teacher
East Valley Middle School
East Helena, Montana

Jane M. Lemons
Science Teacher
Western Rockingham Middle School
Madison, North Carolina

Scott Mandel, Ph.D.
Director and Educational Consultant
Teachers Helping Teachers
Los Angeles, California

Thomas Manerchia
Former Biology and Life Science Teacher
Archmere Academy
Claymont, Delaware

Maurine O. Marchani
Science Teacher and Dept. Chair
Raymond Park Middle School
Indianapolis, Indiana

Jason P. Marsh
Biology Teacher
Montevideo High School and Montevideo Country School
Montevideo, Minnesota

Edith C. McAlanis
Science Teacher and Dept. Chair
Socorro Middle School
El Paso, Texas

Kevin McCurdy, Ph.D.
Science Teacher
Elmwood Junior High School
Rogers, Arkansas

Kathy McKee
Science Teacher
Hoyt Middle School
Des Moines, Iowa

Acknowledgments continue on page 167.

E Environmental Science

Skills Development

Process Skills

QuickLabs

Chapter Labs

Skills Development *(continued)*

Research and Critical Thinking Skills

Apply

Feature Articles

Connections

Mathematics

To the Student

This book was created to make your science experience interesting, exciting, and fun!

Go for It!

Science is a process of discovery, a trek into the unknown. The skills you develop using *Holt Science & Technology*—such as observing, experimenting, and explaining observations and ideas—are the skills you will need for the future. There is a universe of exploration and discovery awaiting those who accept the challenges of science.

Science & Technology

You see the interaction between science and technology every day. Science makes technology possible. On the other hand, some of the products of technology, such as computers, are used to make further scientific discoveries. In fact, much of the scientific work that is done today has become so technically complicated and expensive that no one person can do it entirely alone. But make no mistake, the creative ideas for even the most highly technical and expensive scientific work still come from individuals.

Activities and Labs

The activities and labs in this book will allow you to make some basic but important scientific discoveries on your own. You can even do some exploring on your own at home! Here's your chance to use your imagination and curiosity as you investigate your world.

Keep a ScienceLog

In this book, you will be asked to keep a type of journal called a ScienceLog to record your thoughts, observations, experiments, and conclusions. As you develop your ScienceLog, you will see your own ideas taking shape over time. You'll have a written record of how your ideas have changed as you learn about and explore interesting topics in science.

Know "What You'll Do"

The "What You'll Do" list at the beginning of each section is your built-in guide to what you need to learn in each chapter. When you can answer the questions in the Section Review and Chapter Review, you know you are ready for a test.

Check Out the Internet

You will see this sciLINKS logo throughout the book. You'll be using *sci*LINKS as your gateway to the Internet. Once you log on to *sci*LINKS using your computer's Internet link, type in the *sci*LINKS address. When asked for the keyword code, type in the keyword for that topic. A wealth of resources is now at your disposal to help you learn more about that topic.

In addition to *sci*LINKS you can log on to some other great resources to go with your text. The addresses shown below will take you to the home page of each site.

internet**connect**

This textbook contains the following on-line resources to help you make the most of your science experience.

Visit **go.hrw.com** for extra help and study aids matched to your textbook. Just type in the keyword HST HOME.

Visit **www.scilinks.org** to find resources specific to topics in your textbook. Keywords appear throughout your book to take you further.

Smithsonian Institution® Internet Connections

Visit **www.si.edu/hrw** for specifically chosen on-line materials from one of our nation's premier science museums.

Visit **www.cnnfyi.com** for late-breaking news and current events stories selected just for you.

CHAPTER 1

Sections

Interactions of Living Things

WHAT'S FOR DINNER?

Look at the harsh, frozen environment surrounding this colony of penguins. What do you suppose the penguins eat? Penguin colonies sometimes number more than a million. They survive by swimming underwater to catch and eat krill, a type of tiny shellfish. Penguins compete with whales—and each other—for the supply of krill. In this chapter, you will learn how living things interact and how energy, in the form of food, travels in pathways through different organism groups.

1. Imagine a deer living in a meadow. What does the deer eat? What eats the deer? When the deer dies what happens to its remains?
2. What is the source of energy for plants?

WHO EATS WHOM?

In this activity, you will learn how organisms interact when finding (or becoming) the next meal.

Procedure

1. On each of four **index cards,** print the name of one of the following organisms: white-tailed deer, turkey vulture, oak tree, and cougar
2. Arrange the cards on your desk in a chain to show who eats whom.
3. List the order of your cards in your ScienceLog.
4. In nature, would you expect to see more cougars, more deer, or more oak trees? Arrange the cards in order of most individuals to fewest.

Analysis

5. What might happen to the other organisms if the oak trees were removed from this group? What might happen if the cougars were removed?
6. Are there any organisms in this group that eat more than one kind of food? (Hint: What else might a deer, a cougar, or a turkey vulture eat?) How could you change the order of your cards to show this information? How could you use pieces of string to show these relationships?

SECTION 1

READING WARM-UP

Terms to Learn

ecology
biotic
abiotic
population
community
ecosystem
biosphere

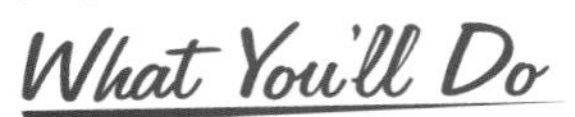

- Distinguish between the biotic and abiotic environment.
- Explain how populations, communities, ecosystems, and the biosphere are related.
- Explain how the abiotic environment relates to communities.

Everything Is Connected

Look at **Figure 1** below. An alligator drifts in a weedy Florida river, watching a long, thin fish called a gar. The gar swims too close to the alligator. Suddenly, in a rush of snapping jaws and splashing water, the gar becomes a meal for the alligator.

It is clear that these two organisms have just interacted with one another. But organisms have many interactions other than simply "who eats whom." For example, alligators dig underwater holes to escape from the heat. Later, after the alligators abandon these holes, fish and other aquatic organisms live in them when the water level gets low during a drought. Alligators also build nest mounds in which to lay their eggs, and they enlarge these mounds each year. Eventually, the mounds become small islands where trees and other plants grow. Herons, egrets, and other birds build their nests in the trees. It is easy to see that alligators affect many organisms, not just the gars that they eat.

Studying the Web of Life

All living things are connected in a web of life. Scientists who study the connections among living things specialize in the science of ecology. **Ecology** is the study of the interactions between organisms and their environment.

An Environment Has Two Parts An organism's environment is anything that affects the organism. An environment consists of two parts. The **biotic** part of the environment is all of the organisms that live together and interact with one another. The **abiotic** part of the environment includes all of the physical factors—such as water, soil, light, and temperature—that affect organisms living in a particular area.

Take another look at Figure 1. How many biotic parts and abiotic parts can you see?

Figure 1 *The alligator affects, and is affected by, many organisms in its environment.*

Organization in the Environment At first glance, the environment may seem disorganized. To ecologists, however, the environment can be arranged into different levels, as shown in **Figure 2.** The first level contains the individual organism. The second level contains similar organisms, forming a population. The third contains different populations, forming a community. The fourth contains a community and its abiotic environment, forming an ecosystem. Finally, the fifth level contains all ecosystems, forming the biosphere. Turn the page and examine **Figure 3** to see these levels in a salt marsh.

Figure 2 **The Five Levels of Environmental Organization**

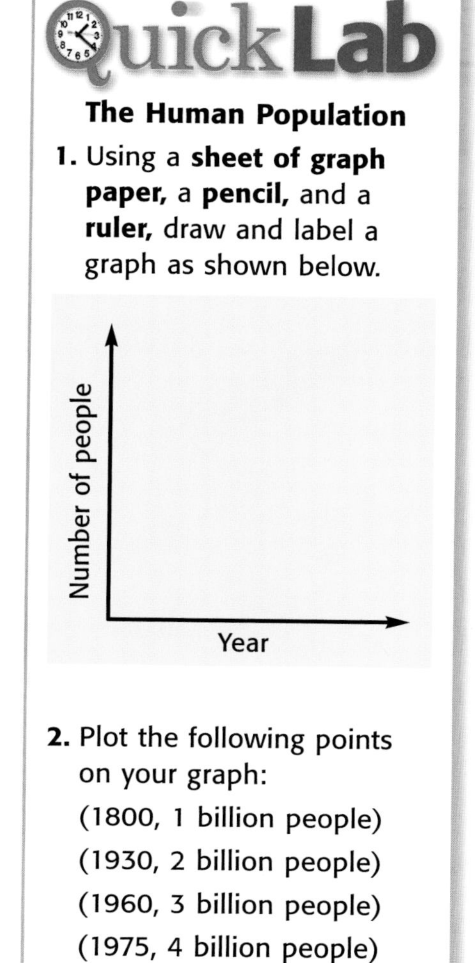

The Human Population

1. Using a **sheet of graph paper,** a **pencil,** and a **ruler,** draw and label a graph as shown below.
2. Plot the following points on your graph:
   (1800, 1 billion people)
   (1930, 2 billion people)
   (1960, 3 billion people)
   (1975, 4 billion people)
   (1987, 5 billion people)
   (1999, 6 billion people)
3. Draw a line connecting the points.
4. Answer the following questions in your ScienceLog.
   a. What does the curve that you have drawn indicate about human population growth?
   b. Do you think the human population can continue to grow indefinitely? Why or why not?

TRY at HOME

Populations A salt marsh is a coastal area where grasslike plants grow. A **population** is a group of individuals of the same species that live together in the same area at the same time. For example, all of the seaside sparrows that live together in a salt marsh are members of a population. The individuals in the population compete with one another for food, nesting space, and mates.

Figure 3 Examine the picture of a salt marsh below. See if you can find examples of each level of organization in this environment.

Communities A **community** consists of all of the populations of different species that live and interact in an area. The various animals and plants you see below form a salt-marsh community. The different populations in a community depend on each other for food, shelter, and many other things.

Ecosystems An **ecosystem** is made up of a community of organisms and its abiotic environment. An ecologist studying the salt-marsh ecosystem would examine how the ecosystem's organisms interact with each other and how temperature, precipitation, and soil characteristics affect the organisms. For example, the rivers and streams that empty into the salt marsh carry nutrients, such as nitrogen, from the land. These nutrients influence how the cordgrass and algae grow.

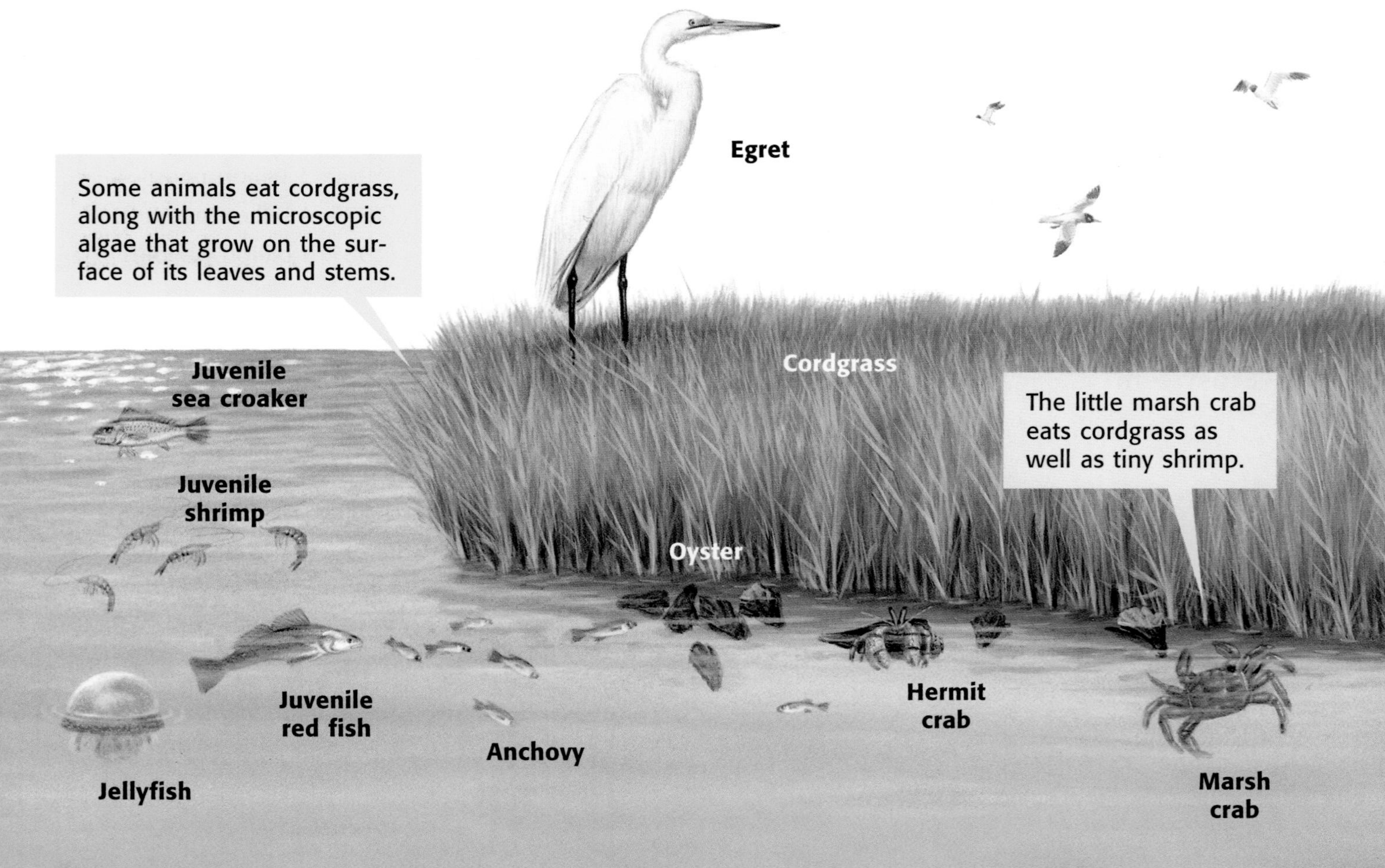

The Biosphere The **biosphere** is the part of Earth where life exists. It extends from the deepest parts of the ocean to very high in the atmosphere, where tiny insects and plant spores drift, and it includes every ecosystem. Ecologists study the biosphere to learn how organisms interact with the abiotic environment—Earth's gaseous atmosphere, water, soil, and rock. The water in the abiotic environment includes both fresh water and salt water as well as water that is frozen in polar icecaps and glaciers.

SECTION REVIEW

1. What is ecology?
2. Give two examples of biotic and abiotic factors in the salt-marsh ecosystem.
3. Using the salt-marsh example, distinguish between populations, communities, ecosystems, and the biosphere.
4. **Analyzing Relationships** What do you think would happen to the other organisms in the salt-marsh ecosystem if the cordgrass were to suddenly die?

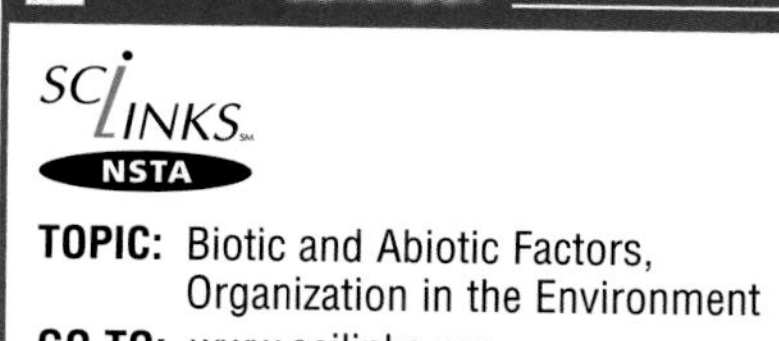

SECTION 2

READING WARM-UP

Living Things Need Energy

Terms to Learn

herbivore
carnivore
omnivore
scavenger
food chain
food web
energy pyramid
habitat
niche

What You'll Do

- Describe the functions of producers, consumers, and decomposers in an ecosystem.
- Distinguish between a food chain and a food web.
- Explain how energy flows through a food web.
- Distinguish between an organism's habitat and its niche.

All living things need energy to survive. For example, black-tailed prairie dogs, which live in the grasslands of North America, eat grass and seeds to get the energy they need. They use this energy to grow, move, heal injuries, and reproduce. In fact, everything a prairie dog does requires energy. The same is true for the plants that grow in the grasslands where the prairie dogs live. Coyotes that stalk prairie dogs, as well as the bacteria and fungi that live in the soil, all need energy.

The Energy Connection

Organisms in a prairie or any community can be divided into three groups based on how they obtain energy. These groups are producers, consumers, and decomposers. Examine **Figure 4** to see how energy passes through these groups in an ecosystem.

Producers Organisms that use sunlight directly to make food are called *producers*. They do this using a process called photosynthesis. Most producers are plants, but algae and some bacteria are also producers. Grasses are the main producers in a prairie ecosystem. Examples of producers in other ecosystems include cordgrass and algae in a salt marsh and trees in a forest. Algae are the main producers in the ocean.

Figure 4 *Follow the pathway of energy as it moves from the sun through the ecosystem.*

Consumers Organisms that eat producers or other organisms for energy are called *consumers*. They cannot use the sun's energy directly like producers can. Instead, consumers must eat producers or other animals to obtain energy. There are several kinds of consumers. A **herbivore** is a consumer that eats plants. Herbivores in the prairie ecosystem include grasshoppers, gophers, prairie dogs, bison, and pronghorn antelope. A **carnivore** is a consumer that eats animals. Carnivores in the prairie ecosystem include coyotes, hawks, badgers, and owls. Consumers known as **omnivores** eat a variety of organisms, both plants and animals. The grasshopper mouse is an example of an omnivore in the prairie ecosystem. It eats insects, scorpions, lizards, and grass seeds. **Scavengers** are animals that feed on the bodies of dead animals. The turkey vulture is a scavenger in the prairie ecosystem. Examples of scavengers in aquatic ecosystems include crayfish, snails, clams, worms, and crabs.

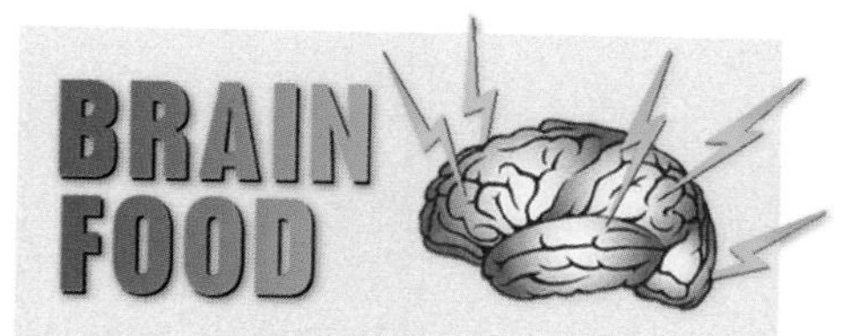

Prairie dogs are not really dogs. They are rodents. They are called dogs because their warning calls sound like the barking of dogs.

Decomposers Organisms that get energy by breaking down the remains of dead organisms are called *decomposers*. Bacteria and fungi are examples of decomposers. These organisms extract the last bit of energy from dead organisms and produce simpler materials, such as water and carbon dioxide. These materials can then be reused by plants and other living things. Decomposers are an essential part of any ecosystem because they are nature's recyclers.

Self-Check

Are you a herbivore, a carnivore, or an omnivore? Explain. *(See page 168 to check your answer.)*

Consumer
A turkey vulture may eat some of the coyote's leftovers. A scavenger can pick bones completely clean.

Decomposer
Any prairie dog remains not eaten by the coyote or the turkey vulture are broken down by bacteria and fungi that live in the soil.

Food Chains and Food Webs

> **Self-Check**
>
> How is a food web different from a food chain? *(See page 168 to check your answer.)*

Figure 4, on pages 8–9, shows a **food chain,** which represents how the energy in food molecules flows from one organism to the next. But because few organisms eat just one kind of organism, simple food chains rarely occur in nature. The many energy pathways possible are more accurately shown by a **food web. Figure 5** shows a simple food web for a woodland ecosystem.

Find the fox and the rabbit in the figure below. Notice that the arrow goes from the rabbit to the fox, showing that the rabbit is food for the fox. The rabbit is also food for the owl. Neither the fox nor the owl is ever food for the rabbit. Energy moves from one organism to the next in a one-way direction, even in a food web. Any energy not immediately used by an organism is stored in its tissues. Only the energy stored in an organism's tissues can be used by the next consumer.

Figure 5 *Energy moves through an ecosystem in complex ways. Most consumers eat a variety of foods and can be eaten by a variety of other consumers.*

Energy Pyramids

A grass plant uses most of the energy it obtains from the sun for its own life processes. But some of the energy is stored in its tissues and is left over for prairie dogs and other animals that eat the grass. Prairie dogs need a lot of energy and have to eat a lot of grass. Each prairie dog uses most of the energy it obtains from eating grass and stores only a little of it in its tissues. Coyotes need even more energy than prairie dogs, so they must eat many prairie dogs to survive. There must be many more prairie dogs in the community than there are coyotes that eat prairie dogs.

The loss of energy at each level of the food chain can be represented by an **energy pyramid,** as shown in **Figure 6.** You can see that the energy pyramid has a large base and becomes smaller at the top. The amount of available energy is reduced at higher levels because most of the energy is either used by the organism or given off as heat. Only energy stored in the tissues of an organism can be transferred to the next level.

Energy Pyramids

Draw an energy pyramid for a river ecosystem that contains four levels—aquatic plants, insect larvae, bluegill fish, and a largemouth bass. The plants obtain 10,000 units of energy from the sun. If each level uses 90 percent of the energy it receives from the previous level, how many units of energy are available to the bass?

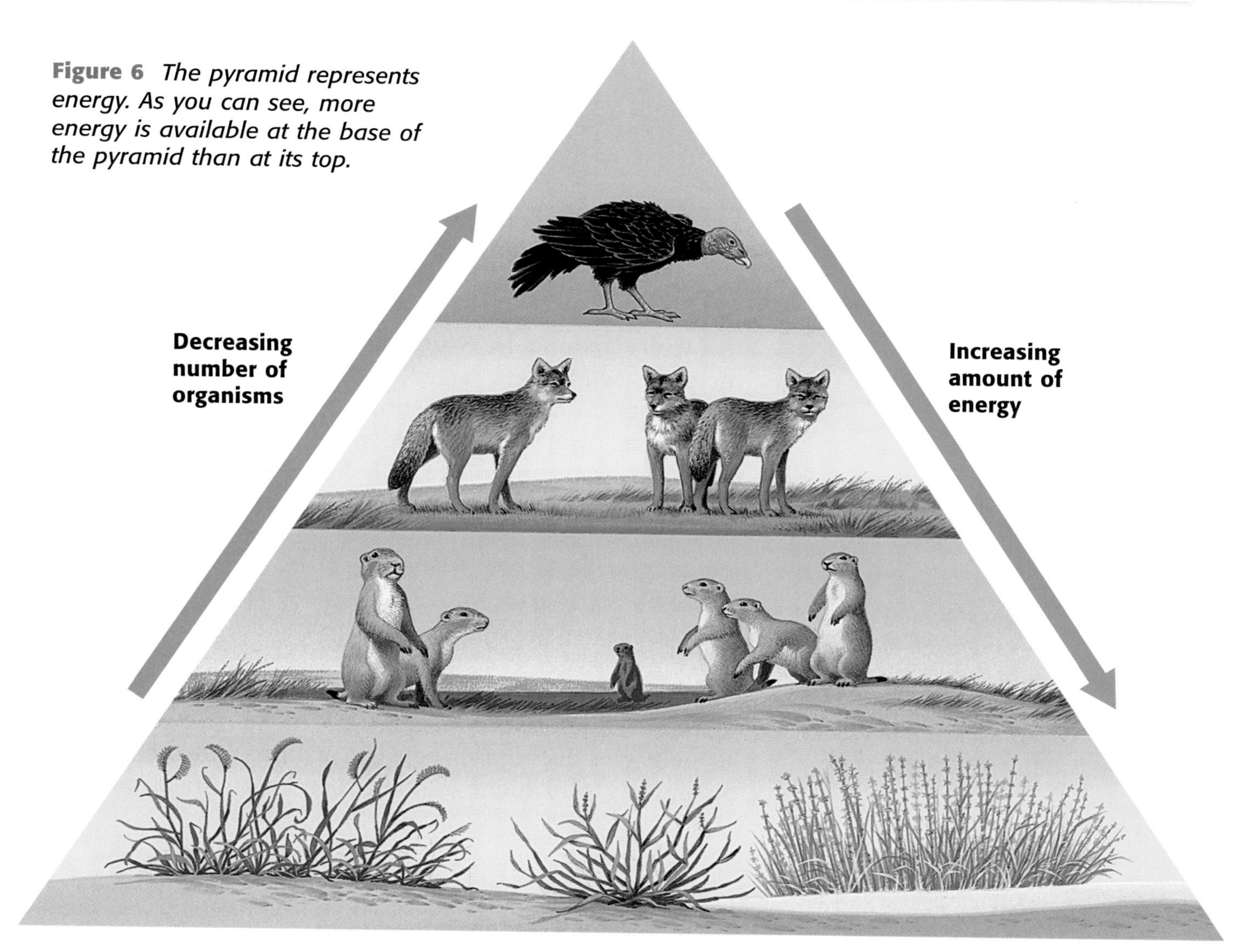

Figure 6 *The pyramid represents energy. As you can see, more energy is available at the base of the pyramid than at its top.*

Figure 7 *As the wilderness was settled, the gray wolf population in the United States declined.*

Wolves and the Energy Pyramid

A single species can be very important to the flow of energy in an environment. Gray wolves, for example, are a consumer species that can control the populations of many other species. The diet of gray wolves can include anything from a lizard to an elk.

Once common throughout much of the United States, gray wolves were almost wiped out as the wilderness was settled. You can see a pair of gray wolves in **Figure 7.** Without wolves, certain other species, such as elk, were no longer controlled. The overpopulation of elk in some areas led to overgrazing and starvation.

Gray wolves were recently restored to the United States at Yellowstone National Park, as shown in **Figure 8.** The U.S. Fish and Wildlife Service hopes this action will restore the natural energy flow in this wilderness area. Not everyone approves, however. Ranchers near Yellowstone are concerned about the safety of their livestock.

Figure 8 *Members of the U.S. Fish and Wildlife Service are moving a caged wolf to a location in Yellowstone National Park.*

Habitat and Niche

An organism's **habitat** is the environment in which it lives. The wolf's habitat was originally very extensive. It included forests, grasslands, deserts, and the northern tundra. Today the wolf's habitat in North America is much smaller. It includes wilderness areas in Montana, Washington, Minnesota, Michigan, Wisconsin, and Canada.

An organism's way of life within an ecosystem is its **niche.** An organism's niche includes its habitat, its food, its predators, and the organisms with which it competes. An organism's niche also includes how the organism affects and is affected by abiotic factors in its environment, such as temperature, light, and moisture.

The Niche of the Gray Wolf

A complete description of a species' niche is very complex. To help you distinguish between habitat and niche, parts of the gray wolf niche are described on the next page.

Gray Wolves Are Consumers Wolves are carnivores. Their diet includes large animals, such as deer, moose, reindeer, sheep, and elk, as well as small animals, such as birds, lizards, snakes, and fish.

Gray Wolves Have a Social Structure Wolves live and hunt in packs, which are groups of about six animals that are usually members of the same family. Each member of the pack has a particular rank within the pack. The pack has two leaders that help defend the pack against enemies, such as other wolf packs or bears.

Figure 9 *In small wolf packs, only the alpha female has pups. They are well cared for, however, by all of the males and females in the pack.*

Gray Wolves Nurture and Teach Their Young A female wolf, shown in **Figure 9,** has five to seven pups and nurses her babies for about 2 months. The entire pack help bring the pups food and baby-sit when the parents are away from the den. It takes about 2 years for the young wolves to learn to hunt. At that time, some young wolves leave the pack to find mates and start their own pack.

Gray Wolves Are Needed in the Food Web If wolves become reestablished at Yellowstone National Park, they will reduce the elk population by killing the old, injured, and diseased elk. This in turn will allow more plants to grow, which will allow animals that eat the plants, such as snowshoe hares, and the animals that eat the hares, such as foxes, to increase in number.

SECTION REVIEW

1. How are producers, consumers (herbivores, carnivores, and scavengers), and decomposers linked in a food chain?
2. How do food chains link together to form a food web?
3. Distinguish between an organism's habitat and its niche using the prairie dog as an example.
4. **Applying Concepts** Is it possible for an inverted energy pyramid to exist, as shown in the figure at right? Explain why or why not.

Scavengers
Carnivores
Herbivores
Producers

Types of Interactions

Terms to Learn

carrying capacity
prey
predator
symbiosis
mutualism
commensalism
parasitism
coevolution

What You'll Do

- Distinguish between the two types of competition.
- Give examples of predators and prey.
- Distinguish between mutualism, commensalism, and parasitism.
- Define *coevolution,* and give an example.

Look at the seaweed forest shown in **Figure 10** below. Notice that some types of organisms are more numerous than others. In natural communities, populations of different organisms vary greatly. The interactions between these populations affect the size of each population.

Figure 10 *This seaweed forest is home to a large number of interacting species.*

Interactions with the Environment

Most living things produce more offspring than will survive. A female frog, for example, might lay hundreds of eggs in a small pond. In a few months, the population of frogs in that pond will be about the same as it was the year before. Why won't the pond become overrun with frogs? An organism, such as a frog, interacts with biotic or abiotic factors in its environment that can control the size of its population.

Limiting Factors Populations cannot grow indefinitely because the environment contains only so much food, water, living space, and other needed resources. When one or more of those resources becomes scarce, it is said to be a *limiting factor.* For example, food becomes a limiting factor when a population becomes too large for the amount of food available. Any single resource can be a limiting factor to population size.

Carrying Capacity The largest population that a given environment can support over a long period of time is known as the environment's **carrying capacity.** When a population grows larger than its carrying capacity, limiting factors in the environment cause the population to get smaller. For example, after a very rainy growing season in an environment, plants may produce a large crop of leaves and seeds. This may cause a herbivore population to grow large because of the unlimited food supply. If the next year has less rainfall than usual, there won't be enough food to support the large herbivore population. In this way, a population may temporarily exceed the carrying capacity. But a limiting factor will cause the population to die back. The population will return to a size that the environment can support over a long period of time.

Self-Check

1. Explain how water can limit the growth of a population.
2. Describe how the carrying capacity for deer in a forest ecosystem might be affected by weather.

(See page 168 to check your answers.)

Interactions Among Organisms

Populations contain interacting individuals of a single species, such as a group of rabbits feeding in the same area. Communities contain interacting populations of several species, such as a coral reef community with many species trying to find living space. Ecologists have described four main ways that species and individuals affect each other: competition, predators and prey, certain symbiotic relationships, and coevolution.

Competition

When two or more individuals or populations try to use the same limited resource, such as food, water, shelter, space, or sunlight, it is called *competition.* Because resources are in limited supply in the environment, their use by one individual or population decreases the amount available to other organisms.

Competition can occur among individuals *within* a population. The elks in Yellowstone National Park are herbivores that compete with each other for the same food plants in the park. This is a big problem for this species in winter. Competition can also occur *between* populations of different species. The different species of trees in **Figure 11** are competing with each other for sunlight and space.

Figure 11 *Some of the trees in this forest grow tall in order to reach sunlight, reducing the amount of sunlight available to shorter trees nearby.*

Predators and Prey

Many interactions among species occur because one organism eats another. The organism that is eaten is called the **prey.** The organism that eats the prey is called the **predator.** When a bird eats a worm, the worm is the prey and the bird is the predator.

Predator Adaptations In order to survive, predators must be able to catch their prey. Predators have a wide variety of methods and abilities for doing this. The cheetah, for example, is able to run at great speed to catch its prey. Other predators, such as the goldenrod spider, shown in **Figure 12,** ambush their prey. The goldenrod spider blends in so well with the goldenrod flower that all it has to do is wait for its next insect meal to arrive.

Figure 12 *The goldenrod spider is difficult for its insect prey to see. Can you see it?*

Prey Adaptations Prey organisms have their own methods and abilities to keep from being eaten. Prey are able to run away, stay in groups, or camouflage themselves. Some prey organisms are poisonous to predators. They may advertise their poison with bright colors to warn predators to stay away. The fire salamander, shown in **Figure 13,** sprays a poison that burns. Predators quickly learn to recognize its warning coloration.

Many animals run away from predators. Prairie dogs run to their underground burrows when a predator approaches. Many small fishes, such as anchovies, swim in groups called schools. Antelopes and buffaloes stay in herds. All the eyes, ears, and noses of the individuals in the group are watching, listening, and smelling for predators. This behavior increases the likelihood of spotting a potential predator.

Some prey species hide from predators by using camouflage. Certain insects resemble leaves so closely that you would never guess they are animals.

Figure 13 *Experienced predators know better than to eat the fire salamander! This colorful animal will make an unlucky predator very sick.*

Symbiosis

Some species have very close interactions with other species. **Symbiosis** is a close, long-term association between two or more species. The individuals in a symbiotic relationship can benefit from, be unaffected by, or be harmed by the relationship. Often, one species lives in or on the other species. The thousands of symbiotic relationships that occur in nature are often classified into three groups: mutualism, commensalism, and parasitism.

Mutualism A symbiotic relationship in which both organisms benefit is called **mutualism.** For example, you and a species of bacteria that lives in your intestines benefit each other! The bacteria get a plentiful food supply from you, and in return you get vitamins that the bacteria produce.

Another example of mutualism occurs between coral and algae. The living corals near the surface of the water provide a home for the algae. The algae produce food through photosynthesis that is used by the corals. When a coral dies, its skeleton serves as a foundation for other corals. Over a long period of time, these skeletons build up large, rocklike formations that lie just beneath the surface of warm, sunny seas, as shown in **Figure 14.**

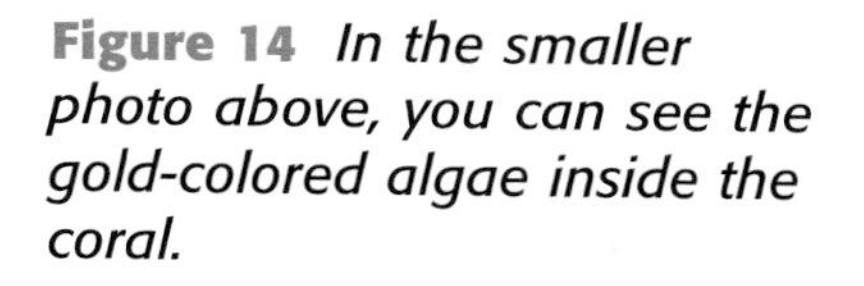

Figure 14 *In the smaller photo above, you can see the gold-colored algae inside the coral.*

Commensalism A symbiotic relationship in which one organism benefits and the other is unaffected is called **commensalism.** One example of commensalism is the relationship between sharks and remoras. **Figure 15** shows a shark with a remora attached to its body. Remoras "hitch a ride" and feed on scraps of food left by sharks. The remoras benefit from this relationship, while sharks are unaffected.

Figure 15 *The remora attached to the shark benefits from the relationship. The shark is neither benefited nor harmed.*

Parasitism A symbiotic association in which one organism benefits while the other is harmed is called **parasitism.** The organism that benefits is called the *parasite.* The organism that is harmed is called the *host.* The parasite gets nourishment from its host, which is weakened in the process. Sometimes a host organism becomes so weak that it dies. Some parasites, such as ticks, live outside the host's body. Other parasites, such as tapeworms, live inside the host's body.

Figure 16 *The tomato hornworm is being parasitized by young wasps. Do you see their cocoons?*

Figure 16 shows a bright green caterpillar called a tomato hornworm. A female wasp laid tiny eggs on the caterpillar. When the eggs hatch, each young wasp will burrow into the caterpillar's body. The young wasps will actually eat the caterpillar alive! In a short time, the caterpillar will be almost completely consumed and will die. When that occurs, the mature wasps will fly away.

In this example of parasitism, the host dies. Most parasites, however, do not kill their hosts. Can you think of reasons why?

Coevolution

Symbiotic relationships and other interactions among organisms in an ecosystem may cause coevolution. **Coevolution** is a long-term change that takes place in two species because of their close interactions with one another.

Coevolution sometimes occurs between herbivores and the plants on which they feed. For example, the ants shown in **Figure 17** have coevolved with a tropical tree called the acacia. The ants protect the tree on which they live by attacking any other herbivore that approaches the tree. The plant has coevolved special structures on its stems that produce food for the ants. The ants live in other structures also made by the tree.

Figure 17 *Ants collect food made by the acacia tree and store the food in their shelter, also made by the tree.*

Coevolution in Australia

In 1859, settlers released 12 rabbits in Australia. There were no predators or parasites to control the rabbit population, and there was plenty of food. The rabbit population increased so fast that the country was soon overrun by rabbits. To control the rabbit population, the Australian government introduced a virus that makes rabbits sick. The first time the virus was used, more than 99 percent of the rabbits died. The survivors reproduced, and the rabbit population grew large again. The second time the virus was used, about 90 percent of the rabbits died. Once again, the rabbit population increased. The third time the virus was used, only about 50 percent of the rabbits died. Suggest what changes might have occurred in the rabbits and the virus.

Coevolution and Flowers Some of the most amazing examples of coevolution are between flowers and their pollinators. (An organism that carries pollen from flower to flower is called a *pollinator.*) When the pollinator travels to the next flower to feed, some of the pollen is left behind on the female part of the flower and more pollen is picked up. Because of pollination, reproduction can take place in the plant. Organisms such as bees, bats, and hummingbirds are attracted to a flower because of its colors, odors, and nectar.

During the course of evolution, hummingbird-pollinated flowers, for example, developed nectar with just the right amount of sugar for their pollinators. The hummingbird's long, thin tongue and beak coevolved to fit into the flowers so that they could reach the nectar. As the hummingbird, like the one shown in **Figure 18,** feeds on the nectar, its head and body become smeared with pollen.

Figure 18 *The bird is attracted to the flower's nectar and picks up the flower's pollen as it feeds.*

SECTION REVIEW

1. Briefly describe one example of a predator-prey relationship. Identify the predator and the prey.
2. Name and define the three kinds of symbiosis.
3. **Analyzing Relationships** Explain the probable relationship between the giant *Rafflesia* flower, which smells like rotting meat, and the carrion flies that buzz around it. HINT: *carrion* means "rotting flesh."

internet**connect**

TOPIC: Producers, Consumers, and Decomposers
GO TO: www.scilinks.org
***sci*LINKS NUMBER:** HSTL440

Making Models Lab

Adaptation: It's a Way of Life

Did you know that organisms have special characteristics called *adaptations* that help them survive changes in their environment? These changes can be climate changes, less food, or disease. These things can cause a population to die out unless some members have adaptations that help them survive. For example, a bird may have an adaptation for eating sunflower seeds and ants. If the ants die out, the bird can still eat seeds in order to live.

In this activity, you will design an organism with special adaptations. Then you will describe how these adaptations help the organism live.

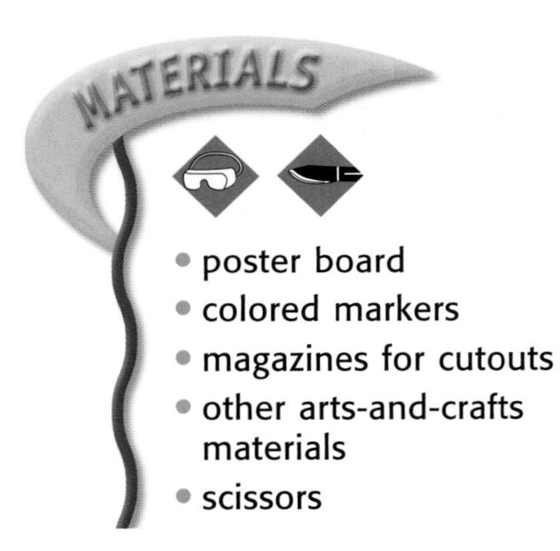

- poster board
- colored markers
- magazines for cutouts
- other arts-and-crafts materials
- scissors

Procedure

1. Study the chart on the next page. Choose one adaptation from each column. For example, an organism might be a scavenger that burrows underground and has spikes on its tail.

2. Design an organism that has the three adaptations you have chosen. Use poster board, colored markers, picture cutouts, construction paper, or any materials of your choice to create your organism.

3. Write a caption on your poster describing your organism. Describe its appearance, where it lives, and how its adaptations help it survive. Give your animal a two-part "scientific" name based on its characteristics.

4. Display your creation in your classroom. Share with classmates how you chose the adaptations for your organism.

Analysis

5. What does your imaginary organism eat?

6. In what environment would your organism be most likely to survive—in the desert, tropical rain forest, plains, icecaps, mountains, or ocean? Explain your answer.

7. What kind of animal is your organism (mammal, insect, reptile, bird, fish)? What modern organism (on Earth today) or ancient organism (extinct) is your imaginary organism most like? Explain the similarities between the two organisms. Do some research outside of class about a real organism that your imaginary organism may resemble.

8. If a sudden climate change occurred, such as daily downpours of rain in a desert, would your imaginary organism survive? What adaptations for surviving such a change does it have?

Adaptations		
Diet	**Type of transportation**	**Special adaptation**
• Carnivore • Herbivore • Omnivore • Scavenger • Decomposer	• Flies • Glides through the air • Burrows underground • Runs fast • Swims • Hops • Walks • Climbs • Floats • Slithers	• Uses sensors to detect heat • Is active only at night and has excellent night vision • Changes color to match its surroundings • Has armor • Has horns • Can withstand extreme temperature changes • Secretes a terrible and sickening scent • Has poison glands • Has specialized front teeth • Has tail spikes • Stores oxygen in its cells so it does not have to breathe continuously • One of your own invention

Chapter Highlights

SECTION 1

Vocabulary

ecology *(p. 4)*
biotic *(p. 4)*
abiotic *(p. 4)*
population *(p. 6)*
community *(p. 6)*
ecosystem *(p. 6)*
biosphere *(p. 7)*

Section Notes

- Ecology is the study of the interactions between organisms and their environment.
- The environment consists of both biotic (living) and abiotic (nonliving) parts.
- A population is a group of the same species living in the same place at the same time. A community is all of the populations of different species living together. An ecosystem is a community and its abiotic environment. The biosphere consists of all of Earth's ecosystems.

Labs

Capturing the Wild Bean *(p. 128)*

SECTION 2

Vocabulary

herbivore *(p. 9)*
carnivore *(p. 9)*
omnivore *(p. 9)*
scavenger *(p. 9)*
food chain *(p. 10)*
food web *(p. 10)*
energy pyramid *(p. 11)*
habitat *(p. 12)*
niche *(p. 12)*

Section Notes

- Organisms that use sunlight directly to make food are called producers. Consumers are organisms that eat other organisms to obtain energy. Decomposers are bacteria and fungi that break down the remains of dead organisms to obtain energy.

Skills Check

Math Concepts

ENERGY PYRAMIDS Try calculating the MathBreak on page 11 as if each unit of energy were $1.00. If you have $10,000.00, but you spend 90 percent, how much do you have left to leave in your will? ($1,000.00) If your heir spends 90 percent of that, how much can your heir leave? ($100.00) After four generations, how much will the inheritance be? ($1.00) Not much, huh? That's why there are very few large organisms at the top of the energy pyramid.

Visual Understanding

FOOD WEBS Several food pathways are shown in the food web in Figure 5 on page 10. However, an actual food web in a woodland ecosystem is much more complex because hundreds of species live there. Find the mouse in Figure 5. How many organisms feed on the mouse? How many organisms feed on the butterfly? What might happen to this ecosystem if these animals were eliminated?

SECTION 2

- A food chain shows how energy flows from one organism to the next.
- Because most organisms eat more than one kind of food, there are many energy pathways possible; these are represented by a food web.
- Energy pyramids demonstrate that most of the energy at each level of the food chain is used up at that level and is unavailable for organisms higher on the food chain.
- An organism's habitat is the environment in which it lives. An organism's niche is its role in the ecosystem.

SECTION 3

Vocabulary

carrying capacity *(p. 15)*
prey *(p. 16)*
predator *(p. 16)*
symbiosis *(p. 17)*
mutualism *(p. 17)*
commensalism *(p. 17)*
parasitism *(p. 18)*
coevolution *(p. 18)*

Section Notes

- Population size changes over time.
- Limiting factors slow the growth of a population. The largest population that an environment can support over a long period of time is called the carrying capacity.
- When one organism eats another, the organism that is eaten is the prey, and the organism that eats the prey is the predator.
- Symbiosis is a close, long-term association between two or more species. There are three general types of symbiosis: mutualism, commensalism, and parasitism.
- Coevolution involves the long-term changes that take place in two species because of their close interactions with one another.

internetconnect

GO TO: go.hrw.com

Visit the **HRW** Web site for a variety of learning tools related to this chapter. Just type in the keyword:

KEYWORD: HSTINT

GO TO: www.scilinks.org

Visit the **National Science Teachers Association** on-line Web site for Internet resources related to this chapter. Just type in the ***sci*LINKS** number for more information about the topic:

TOPIC: Biotic and Abiotic Factors — ***sci*LINKS NUMBER:** HSTL430
TOPIC: Organization in the Environment — ***sci*LINKS NUMBER:** HSTL435
TOPIC: Producers, Consumers, and Decomposers — ***sci*LINKS NUMBER:** HSTL440
TOPIC: Food Chains and Food Webs — ***sci*LINKS NUMBER:** HSTL445
TOPIC: Habitats and Niches — ***sci*LINKS NUMBER:** HSTL450

Chapter Review

USING VOCABULARY

To complete the following sentences, choose the correct term from each pair of terms listed below:

1. An organism's environment has two parts, the __?__, or living, and the __?__, or nonliving. *(biotic* or *abiotic)*
2. A __?__ is a group of individuals of the same species that live in the same area at the same time. *(community* or *population)*
3. A community and its abiotic environment make up a(n) __?__. *(ecosystem* or *food web)*
4. Organisms that use photosynthesis to obtain energy are called __?__. *(producers* or *decomposers)*
5. The environment in which an organism lives is its __?__, and the role the organism plays in an ecosystem is its __?__. *(niche* or *habitat)*

UNDERSTANDING CONCEPTS

Multiple Choice

6. A tick sucks blood from a dog. In this relationship, the tick is the __?__ and the dog is the __?__.
   a. parasite, prey
   b. predator, host
   c. parasite, host
   d. host, parasite
7. Resources such as water, food, or sunlight are more likely to be limiting factors
   a. when population size is decreasing.
   b. when predators eat their prey.
   c. when the population is small.
   d. when a population is approaching the carrying capacity.
8. "Nature's recyclers" are
   a. predators.
   b. decomposers.
   c. producers.
   d. omnivores.
9. A beneficial association between coral and algae is an example of
   a. commensalism.
   b. parasitism.
   c. mutualism.
   d. predation.
10. How energy moves through an ecosystem can be represented by
    a. food chains.
    b. energy pyramids.
    c. food webs.
    d. All of the above
11. The base of an energy pyramid represents which organisms in an ecosystem?
    a. producers
    b. carnivores
    c. herbivores
    d. scavengers
12. Which of the following is the correct order in a food chain?
    a. sun → producers → herbivores → scavengers → carnivores
    b. sun → consumers → predators → parasites → hosts
    c. sun → producers → decomposers → consumers → omnivores
    d. sun → producers → herbivores → carnivores → scavengers
13. Remoras and sharks have a relationship best described as
    a. mutualism.
    b. commensalism.
    c. predator and prey.
    d. parasitism.

Short Answer

14. Briefly describe the habitat and niche of the gray wolf.
15. What might different species of trees in a forest compete for?
16. How do limiting factors affect the carrying capacity of an environment?
17. What is coevolution?

Concept Mapping

18. Use the following terms to create a concept map: individual organisms, producers, populations, ecosystems, consumers, herbivores, communities, carnivores, the biosphere.

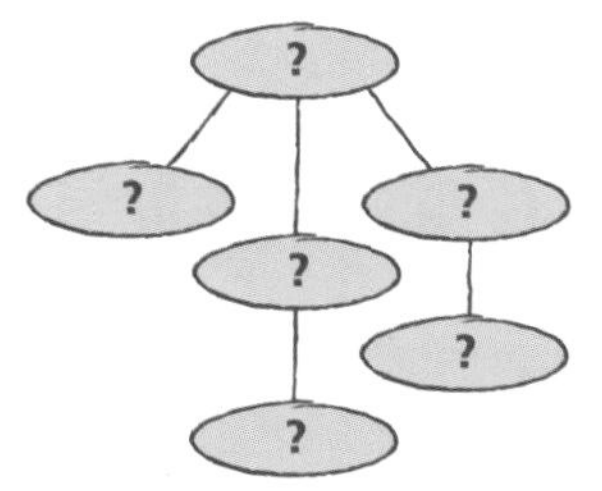

CRITICAL THINKING AND PROBLEM SOLVING

Write one or two sentences to answer the following questions:

19. Could a balanced ecosystem contain producers and consumers but no decomposers? Why or why not?

20. Some biologists think that certain species, such as alligators and wolves, help maintain biological diversity in their ecosystems. Predict what might happen to other species, such as gar fish or herons, if alligators were to become extinct in the Florida Everglades.

21. Does the Earth have a carrying capacity for humans? Explain your answer.

22. Explain why it is important to have a variety of organisms in a community of interacting species. Give an example.

MATH IN SCIENCE

23. The plants in each square meter of an ecosystem obtained 20,810 Calories of the sun's energy by photosynthesis per year. The herbivores in that ecosystem ate all of the plants, but they obtained only 3,370 Calories of energy. How much energy did the plants use for their own life processes?

INTERPRETING GRAPHICS

Examine the following graph, which shows the population growth of a species of *Paramecium,* a slipper-shaped, single-celled microorganism, over a period of 18 days. Food was occasionally added to the test tube in which the paramecia were grown. Answer the following questions:

24. What is the carrying capacity of the test tube as long as food is added?

25. Predict what will happen to the population if the researcher stops adding food to the test tube.

26. What keeps the number of *Paramecium* at a steady level?

27. Predict what might happen if the amount of water is doubled and the food supply stays the same.

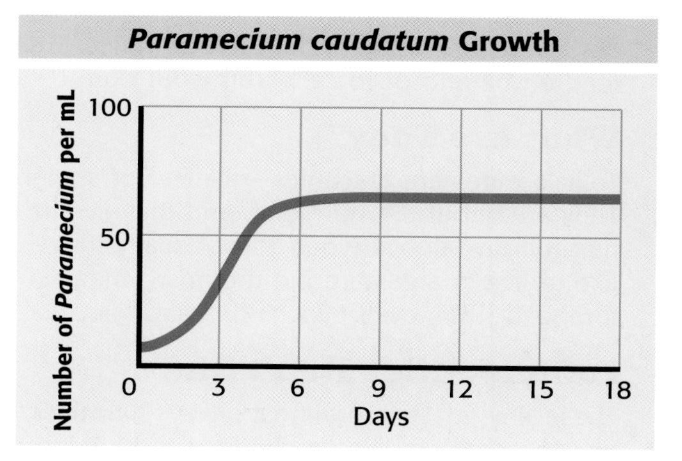

Reading Check-up

Take a minute to review your answers to the Pre-Reading Questions found at the bottom of page 2. Have your answers changed? If necessary, revise your answers based on what you have learned since you began this chapter.

Health Watch

An Unusual Guest

What has a tiny tubelike body and short stumpy legs and lives upside down in your eyebrows and eyelashes? Would you believe a small animal? It's called a follicle mite, and humans are its host organism. Like all large animals, human beings are hosts to a variety of smaller creatures. They live in or on our bodies and share our bodies' resources. But none of our guests are stranger than follicle mites. They feed on oil and dead cells from your skin.

▲ A follicle mite is smaller than the period at the end of this sentence.

What Are They?

Follicle mites are arachnids—relatives of spiders. They are about 0.4 mm long, and they live in hair follicles all over your body. Usually they like to live in areas around the nose, cheek, forehead, chin, eyebrows, and eyelashes.

Follicle Mites Don't Bite

These tiny guests are almost always harmless, and they seldom live on children and adolescents. And you probably wouldn't even know they were there. Studies reveal that between 97 percent and 100 percent of all adults have these mites. Except in rare cases, follicle mites in adults are also pretty harmless.

Some Health Concerns

Although follicle mites rarely cause problems, they are sometimes responsible for an acnelike condition around the nose, eyebrows, and eyelashes. A large number of mites (up to 25) may live in the same follicle. This can cause an inflammation of the follicle. The follicle does not swell like acne; instead it becomes red and itchy.

Mites living in eyelashes and on eyelids can irritate those areas. The inflammation causes itchy eyelids or eyebrows. But such inflammations are rare, and the condition clears up very quickly when suitable medication is applied. So while follicle mites may be one of the strangest guests living on human skin, they are almost never a problem.

Other Companions

Many tiny organisms make their home in humans' bodies. Bacteria within the body may help maintain proper pH levels. Even *Escherichia coli,* a type of bacterium that can cause severe health problems, lives in the human colon. Without *E. coli,* a person would be unable to produce enough vitamin K or folic acid.

On Your Own

▶ Do some more research on follicle mites. Search for *Demodex folliculorum* or *Demodex brevis.* Find out more about some of the other strange organisms that rely on humans' bodies for food and shelter. Report on your findings, or write a story from the organism's point of view.

EYE ON THE ENVIRONMENT

Alien Invasion

A group of tiny aliens left their ship in Mobile, Alabama. Their bodies were red and shiny, and they walked on six legs. The aliens looked around and then quietly crawled off to make homes in the new land.

Westward Ho!

In 1918, fire ants were accidentally imported into the United States by a freighter ship from South America. In the United States, fire ants have no natural predators or competitors. In addition, these ants are extremely aggressive, and their colonies can harbor many queens, instead of just one queen, like many other ant species. With all these advantages, it is not surprising that the ants have spread like wildfire. By 1965, fire-ant mounds were popping up on the southeastern coast and as far west as Texas. Today they are found in at least 10 southern states and may soon reach as far west as California.

▲ Three types of fire ants are found in a colony: the queen, workers, and males. Notice how the queen ant dwarfs the worker ants.

Jaws of Destruction

Imported fire ants have done a lot of damage as they have spread across the United States. Because they are attracted to electrical currents, they chew through wire insulation, causing shorts in electrical circuits. The invaders have also managed to disturb the natural balance of native ecosystems. In some areas, they have killed off 70 percent of the native ant species and 40 percent of other native insect species. Each year, about 25,000 people seek medical attention for painful fire-ant bites.

Fighting Fire

Eighty years after the fire ants' introduction into the United States, the destructive ants continue to multiply. About 157 chemical products, including ammonia, gasoline, extracts from manure, and harsh pesticides, are registered for use against fire ants, but most have little or no success. Unfortunately, many of these remedies also harm the environment. By 1995, the government had approved only one fire-ant bait for large-scale use.

An Ant-Farm Census

▶ How many total offspring does a single fire-ant queen produce if she lives for 5 years and produces 1,000 eggs a day? If a mound contains 300,000 ants, how many mounds will her offspring fill?

CHAPTER 2

Cycles in Nature

Sections

1. What is meant by *recycling*?
2. What happens to rainwater after it falls to Earth?

A CLASSROOM AQUARIUM

Did you know an aquarium is a small environment? In this activity, you will put an aquarium together. As you plan the aquarium, think about how all the parts are connected with each other.

Procedure

1. Get a **tank** from your teacher. Check the Internet, your library, or a pet store to find directions on the proper way to clean and prepare an aquarium.
2. Find out about the kinds of **plants** and **animals** that you can put in your aquarium.
3. Choose a place to put the tank, and tell your teacher your plans. Then set up the aquarium.

Analysis

4. How is the aquarium similar to and different from a natural body of water? Identify the limitations of your model.
5. After you read Section 1, take another look at the aquarium. See if you can name all the parts of this small ecosystem.

Desert Post

Ever tried to send a desert to anyone? Thanks to the U.S. Postal Service, you can! Stamps were made from this picture of the Sonoran Desert. The stamps are intended to promote a greater appreciation of the diversity of the Sonoran Desert. The desert scene was the first of a series that commemorates America's natural environment. In this chapter, you will learn about the natural environment and how it works.

SECTION 1

READING WARM-UP

Terms to Learn

precipitation
evaporation
ground water
decomposition
combustion

What You'll Do

- Trace the cycle of water between the atmosphere, land, and oceans.
- Diagram the carbon cycle, and explain its importance to living things.
- Diagram the nitrogen cycle, and explain its importance to living things.

The Cycles of Matter

The matter in your body has been on Earth since the planet was formed billions of years ago! *Matter,* which is anything that occupies space and has mass, is used over and over again. Each kind of matter has its own cycle. In these cycles, matter moves among the environment and living things.

The Water Cycle

The movement of water among the oceans, atmosphere, land, and living things is known as the *water cycle*. Locate each part of the water cycle in **Figure 1** as it is discussed.

Precipitation Water moves from the atmosphere to the land and oceans as **precipitation,** which includes rain, snow, sleet, and hail. About 91 percent of precipitation falls into the ocean. The rest falls on land, renewing the supply of fresh water.

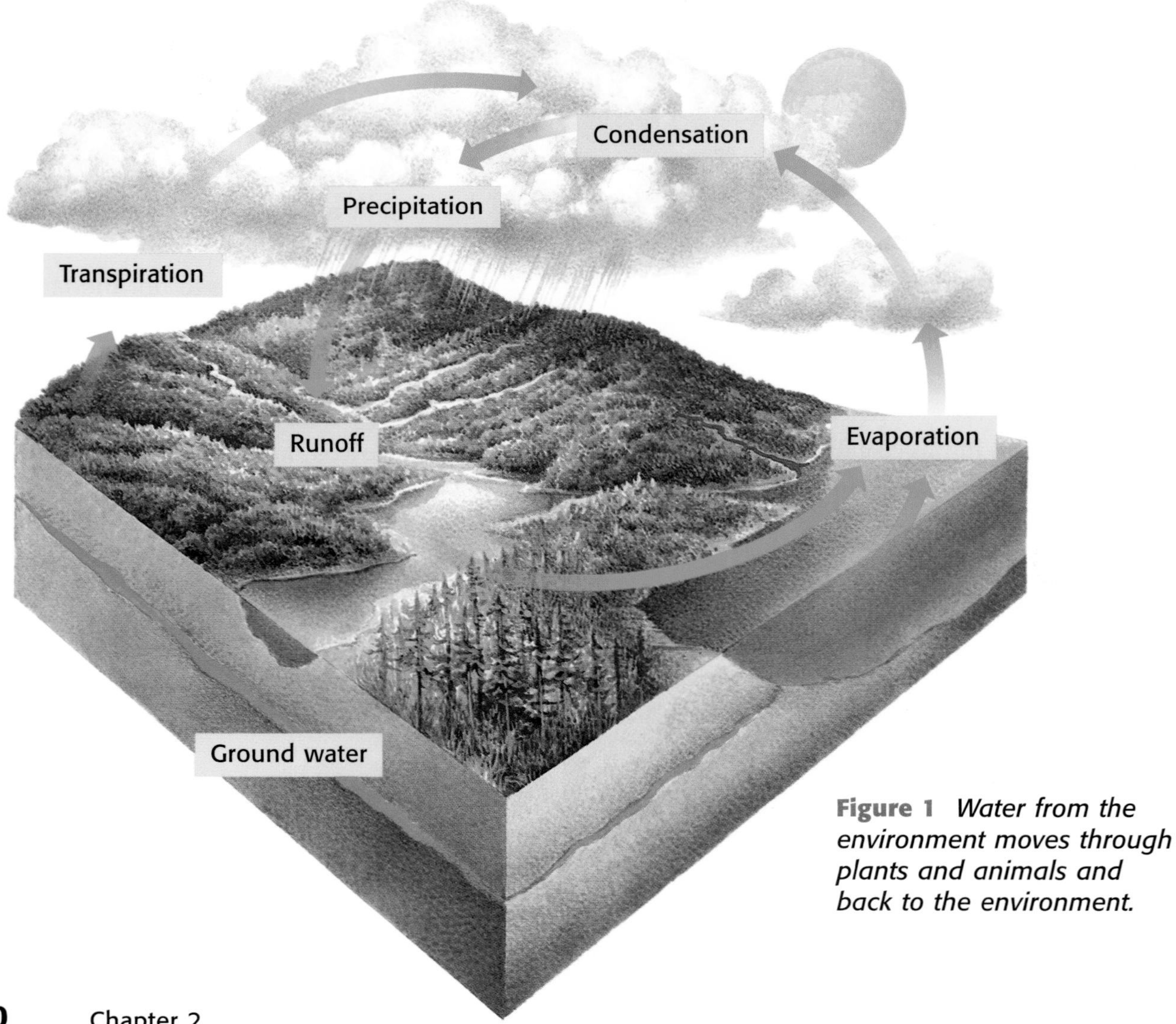

Figure 1 *Water from the environment moves through plants and animals and back to the environment.*

Evaporation Water cycles back to the atmosphere through evaporation. During **evaporation,** the sun's heat causes water to change from liquid to vapor. When the water vapor cools during the process of *condensation,* it forms a liquid that can fall to the Earth as precipitation.

Ground Water Some precipitation seeps into the ground, where it is stored in underground caverns or in porous rock. This water, known as **ground water,** may stay in the ground for hundreds or even thousands of years. Ground water provides water to the soil, streams, rivers, and oceans.

Water and Life All organisms, from tiny bacteria to animals and plants, contain a lot of water. Your body is composed of about 70 percent water. Water carries waste products away from body tissues. Water also helps regulate body temperature through perspiration and evaporation, returning water to the environment in a process called *transpiration.* Without water, there would be no life on Earth.

Carbon dioxide is being released into the atmosphere in increasing quantities. Carbon dioxide causes the atmosphere to hold heat. The warmer atmosphere causes the temperatures of the land and ocean to rise. This is known as global warming.

The Carbon Cycle

Carbon is essential to living things because it is part of all biological molecules. The movement of carbon from the environment into living things and back into the environment is known as the *carbon cycle,* shown in **Figure 2.**

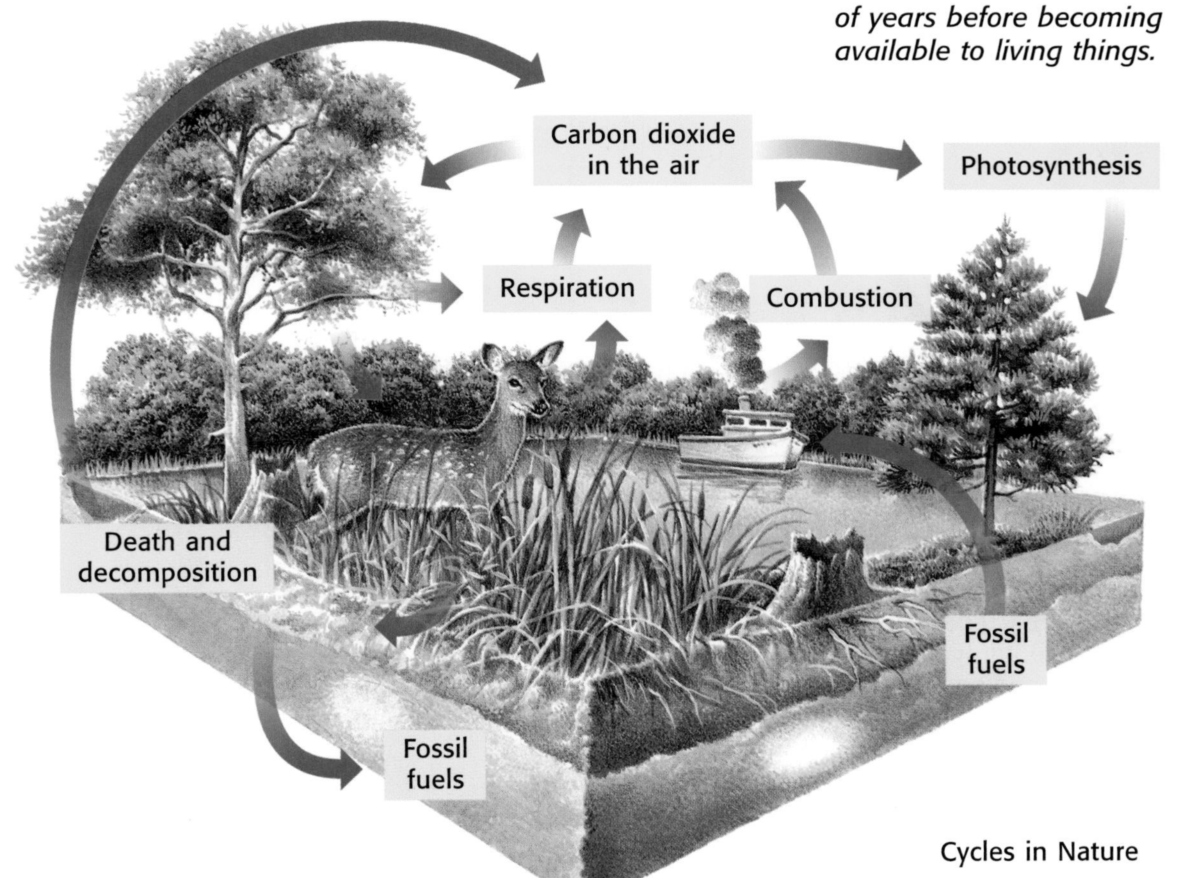

Figure 2 *Carbon may remain in the environment for millions of years before becoming available to living things.*

Combustion

Place a **candle** on a **jar lid** and secure it with **modeling clay.** Light the candle. Hold the **jar** very close to the candle flame. What is deposited on the jar? Where did the substance come from? Now place the jar over the candle. What is deposited inside the jar? Where did this substance come from?

Photosynthesis Photosynthesis is the process by which carbon cycles from the environment into living things. During photosynthesis, plants use carbon dioxide from the air to make sugars. Most animals get the carbon they need by eating plants.

Respiration How does carbon return to the environment? Animals and plants both respire. During *respiration,* sugar molecules are broken down to release energy. Carbon dioxide and water are released as byproducts.

Decomposition The breakdown of dead materials into carbon dioxide and water is called **decomposition.** When fungi and bacteria decompose organic matter, they return carbon to the environment.

Combustion The carbon in coal, oil, and natural gas returns to the atmosphere as carbon dioxide when these fuels are burned. The process of burning fuel is known as **combustion.** Combustion provides much of the fuel people need to drive cars, heat homes, and make electricity.

The Nitrogen Cycle

The movement of nitrogen from the environment to living things and back again is called the *nitrogen cycle,* shown in **Figure 3.**

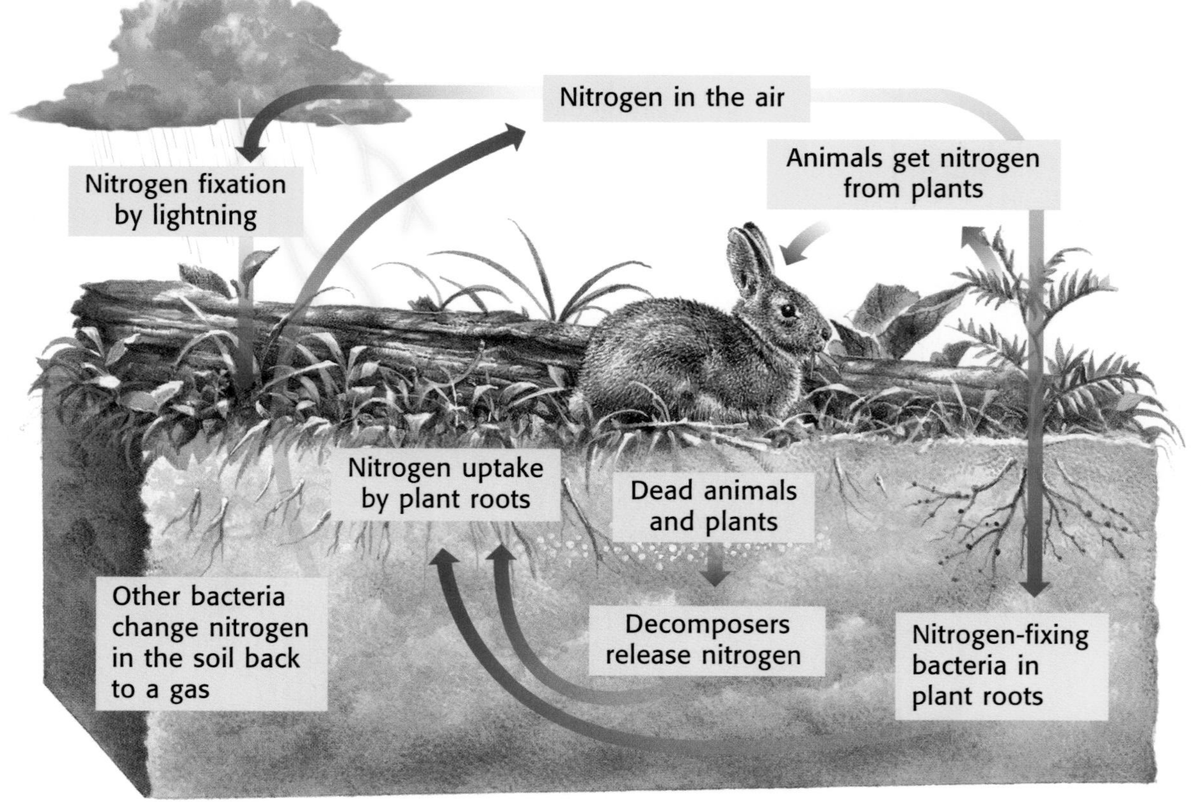

Figure 3 *Without bacteria, nitrogen could not enter living things or be returned to the atmosphere.*

A Sea of Nitrogen About 78 percent of the Earth's atmosphere is nitrogen gas. However, most organisms cannot use nitrogen gas to obtain the nitrogen they need to build proteins and DNA. But bacteria in the soil are able to change nitrogen gas into forms that can be used by plants. This is called *nitrogen fixation.* Most animals get the nitrogen they need by eating plants.

Back to Gas The final step of the nitrogen cycle is also performed by bacteria in the soil. These bacteria are different species than the bacteria that fix nitrogen. The bacteria break down dead organisms and animal wastes. This process produces nitrogen gas, which is returned to the atmosphere.

Gallons Galore

An average person in the United States uses about 78 gal of water each day. How many liters is this? How many cubic centimeters? Remember: 1 gal = 3.79 L and 1 mL = 1 cm^3.

The Pollution Cycle

Isabel read an article about how power plants near her home emit sulfur dioxide into the atmosphere. When sulfur dioxide mixes with water, it forms sulfuric acid, which is extremely toxic to living things. Isabel also learned that sulfur dioxide from these power plants has killed all the fish in a lake hundreds of kilometers away. Trees growing near the lake were also killed. Using what you know about the water cycle, write a letter to Isabel explaining how this could happen.

SECTION REVIEW

1. How are precipitation, evaporation, and ground water involved in the water cycle?
2. Draw a simple diagram of the nitrogen cycle. Make sure you include how animals get nitrogen.
3. **Analyzing Relationships** How is decomposition related to the carbon cycle?

internetconnect

SCI LINKS NSTA

TOPIC: The Water Cycle, The Nitrogen Cycle
GO TO: www.scilinks.org
***sci*LINKS NUMBER:** HSTL455, HSTL465

Ecological Succession

Terms to Learn

succession
pioneer species

What You'll Do

- Define *succession.*
- Contrast primary and secondary succession.

Imagine you have a time machine that can take you back to the summer of 1988. If you had visited Yellowstone National Park during that year, you would have found large areas of the park burned to the ground. When the fires were put out, a layer of gray ash blanketed the forest floor. Most of the trees were dead, although many of them were still standing, as shown in **Figure 4.**

Figure 4 *Parts of Yellowstone National Park burned in 1988.*

Figure 5 *In the spring of 1989, regrowth was evident in the burned parts of Yellowstone National Park.*

Regrowth of a Forest

The following spring, the appearance of the "dead" forest began to change. In **Figure 5,** you can see that some of the dead trees are beginning to fall over, and small, green plants have begun to grow in large numbers. National Park foresters report that the number and kinds of plants growing in the recovering area have increased each year since the fire.

A gradual development of a community over time, such as the regrowth of the burned areas of Yellowstone National Park, is called **succession.** Succession takes place in all communities, not just those affected by disturbances such as forest fires. Succession occurs through predictable stages over time, as described on the following pages.

Primary Succession

Sometimes a small community of living things starts to live in an area that did not previously contain any plants or other organisms. There is no soil in this area, usually just bare rock. Over a very long time, a series of organisms live and die on the rock, and the rock is slowly transformed into soil. This process is called *primary succession.*

1 A slowly retreating glacier exposes bare rock where nothing lives, and primary succession begins.

2 Most primary succession begins with lichens. Acids from the lichens begin breaking the rocks into small particles. These particles mix with the remains of dead lichens to start forming soil. Because lichens are the first organisms to live on the rock, they are called **pioneer species.**

3 After many years, the soil is deep enough for mosses to grow. The mosses eventually replace the lichens. Other tiny organisms, such as insects, also make their home among the lichens and mosses. When they die, their remains add to the soil.

4 Over time, the soil layer thickens, and the moss community is replaced by ferns. The ferns in turn may be replaced by grasses and wild-flowers. Once there is sufficient soil, shrubs and small trees come into the area.

5 After hundreds or even thousands of years, the soil may be deep enough to support a forest.

Secondary Succession

Sometimes an existing community is destroyed by a natural disaster, such as fire or flood. Or, a farmer might stop growing crops in an area that had been cleared. In either case, if soil is left intact, the original plant community may regrow through a series of stages called *secondary succession.*

1 The first year after a farmer stops growing crops, or after some other major disturbance, many weeds grow. Crabgrass is usually the most common weed during the first year.

2 By the second year, new weedy plants appear. Their seeds may have blown into the field by the wind, or insects may have carried them. One of the most common weeds during the second year is horseweed.

3 In 5 to 15 years, small pine trees may start growing among the weeds. The pines continue to grow, and, after about 100 years, a forest may form.

4 As older pines die, they may be replaced by hardwoods if the climate can support them.

Self-Check

Describe the differences between primary succession and secondary succession. *(See page 168 to check your answer.)*

Where Does It All End? In the early stages of succession only a few species grow in an area. These species grow fast and make many seeds that scatter easily. Because there are only a few species, they are open to invasion by other, longer-lasting species, disease, and other disturbances. In later stages of succession there are usually many more species present. Because of this, there are more pathways available to absorb disturbances. For example, in a mature forest, many species will survive an invasion by insects if these insects prefer to eat only one species of plant.

Eventually, if an area experiences no fires or other disturbances, it will reach a more or less stable stage. Communities change over time even though they are considered to be stable. A stable community may not always be a hardwood forest. Look at **Figure 6.** Why might a stable hardwood forest not develop there? The answer is that the area does not have the kind of climate that will support a stable hardwood forest. The climate in this area supports a desert community.

Make a diorama of the stages of primary succession. Use boxes, craft supplies, rocks, twigs, and other materials that you can find at home or at school. You can use one large box showing all the stages of primary succession, or you might use several small boxes, each showing a single stage of primary succession. Label and explain each stage.

Figure 6 *This is how a stable community in the Sonoran Desert in Arizona looks in spring.*

SECTION REVIEW

1. Define *succession.*
2. Describe succession in an abandoned field.
3. **Applying Concepts** Explain why soil formation is always the first stage of primary succession. Does soil formation stop when trees begin to grow? Why or why not?

Making Models Lab

A Passel o' Pioneers

Succession is the progressive replacement of one type of community by another in a single area. The area could be one that has never seen life before and has no soil, such as a cooled lava flow or a rock uncovered by a retreating glacier. In an area where there is no soil, the process is called primary succession. In an area where soil is already there, such as a forest after a fire, the process is called secondary succession. In this exercise, you will build a model of secondary succession using natural soil.

MATERIALS

- large fishbowl
- 500 g of soil from home or schoolyard
- balance
- 250 mL graduated cylinder
- plastic wrap
- water
- protective gloves

Procedure

1. Using a balance, measure 500 g of the soil you brought from home or the schoolyard. Place the soil into the fishbowl. Wet the soil with 250 mL of water. Cover the top of the fishbowl with plastic wrap, and place the fishbowl in a sunny window.
   Caution: Do not touch your face, eyes, or mouth during this exercise. Wash your hands when you are finished.
2. For two weeks, watch the soil for any new growth. Describe and draw any new plants you see. Record these and all other observations in your ScienceLog.
3. Name and record as many of these new plants as you can.

Analysis

4. What kinds of plants grew in your model of secondary succession? Were they tree seedlings, grass, or weeds?

5. Were the plants that sprouted in the fishbowl strange or ordinary for your area?

6. Explain how the plants that grew in your model of secondary succession can be called pioneer species.

7. Using your observations, explain how ecological succession worked to maintain equilibrium in your model.

Going Further

Look at each picture on this page. Analyze whether each area, if left alone, would go through primary or secondary succession. You may decide that an area will not go through succession at all. Explain your reasoning.

A pond choked with vegetation

Bulldozed land

Mount St. Helens volcano

Chapter Highlights

SECTION 1

Vocabulary

precipitation *(p. 30)*
evaporation *(p. 31)*
ground water *(p. 31)*
decomposition *(p. 32)*
combustion *(p. 32)*

Section Notes

- Materials used by living things continually cycle through ecosystems.
- In the water cycle, water moves through the ocean, atmosphere, land, and living things.
- Precipitation, evaporation, transpiration, and condensation are important processes in the water cycle.
- Water that falls is held in soil or porous rocks as ground water.
- Photosynthesis, respiration, decomposition, and combustion are important steps in the carbon cycle.
- Carbon enters plants from the nonliving environment as carbon dioxide.
- The process of changing nitrogen gas into forms that plants can use is called nitrogen fixation.

Labs

Nitrogen Needs *(p. 130)*

Skills Check

Math Concepts

SAVING WATER Flushing the toilet accounts for almost half the water a person uses in a day. Some toilets use up to 6 gal per flush. More-efficient toilets use about 1.5 gal per flush. How many liters of water can you save using a more-efficient toilet if you flush five times a day?

$$6 \text{ gal} - 1.5 \text{ gal} = 4.5 \text{ gal}$$
$$4.5 \text{ gal} \times 5 \text{ flushes} = 22.5 \text{ gal}$$
$$1 \text{ gal is equal to } 3.79 \text{ L}$$
$$3.79 \text{ L} \times 22.5 \text{ gal} = 85.275 \text{ L of water saved}$$

Visual Understanding

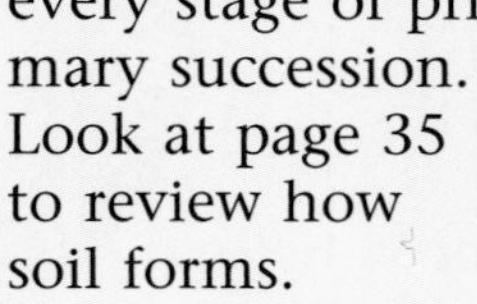

SOIL FORMATION The formation of soil is part of every stage of primary succession. Look at page 35 to review how soil forms.

SECTION 2

Vocabulary

succession *(p. 34)*
pioneer species *(p. 35)*

Section Notes

- Ecological succession is the gradual development of communities over time. Often a series of stages is observed during succession.
- Primary succession occurs in an area that was not previously inhabited by living things; no soil is present.
- Secondary succession occurs in an area where an earlier community was disturbed by fire, landslides, floods, or plowing for crops; soil is present.

internet connect

GO TO: go.hrw.com

Visit the **HRW** Web site for a variety of learning tools related to this chapter. Just type in the keyword:

KEYWORD: HSTCYC

GO TO: www.scilinks.org

Visit the **National Science Teachers Association** on-line Web site for Internet resources related to this chapter. Just type in the ***sci*LINKS** number for more information about the topic:

TOPIC: The Water Cycle — ***sci*LINKS NUMBER:** HSTL455
TOPIC: The Carbon Cycle — ***sci*LINKS NUMBER:** HSTL460
TOPIC: The Nitrogen Cycle — ***sci*LINKS NUMBER:** HSTL465
TOPIC: Succession — ***sci*LINKS NUMBER:** HSTL470

Chapter Review

USING VOCABULARY

To complete the following sentences, choose the correct term from each pair of terms listed below:

1. During __?__, water moves from the atmosphere to the land and ocean. *(evaporation* or *precipitation)*
2. All biological molecules contain __?__. *(carbon* or *carbon dioxide)*
3. The combustion of coal, oil, and natural gas is part of the __?__. *(nitrogen cycle* or *carbon cycle)*
4. The development of a community on bare, exposed rock is an example of __?__. *(primary succession* or *secondary succession)*
5. The recovery of Yellowstone National Park following the fires of 1988 is an example of __?__. *(primary succession* or *secondary succession)*

UNDERSTANDING CONCEPTS

Multiple Choice

6. Water changes from a liquid to a vapor during
   a. precipitation.
   b. respiration.
   c. evaporation.
   d. decomposition.

7. The process of burning fuel, such as oil and coal, is
   a. combustion.
   b. respiration.
   c. decomposition.
   d. photosynthesis.

8. One of the most common plants in a recently abandoned farm field is
   a. horseweed.
   b. young pine trees.
   c. young oak and hickory trees.
   d. crabgrass.

9. Which of the following statements about ground water is true?
   a. It stays underground for a few days.
   b. It is stored in underground caverns or porous rock.
   c. It is salty like ocean water.
   d. It never reenters the water cycle.

10. Which of the following processes produces carbon dioxide?
    a. decomposition
    b. respiration
    c. combustion
    d. all of the above

11. During nitrogen fixation, nitrogen gas is converted into a form that __?__ can use.
    a. plants
    b. animals
    c. fungi
    d. all of the above

12. Bacteria are essential to
    a. combustion.
    b. photosynthesis.
    c. nitrogen fixation.
    d. evaporation.

13. The pioneer species on bare rock are usually
    a. ferns.
    b. pine trees.
    c. mosses.
    d. lichens.

Short Answer

14. Is snow a part of the water cycle? Why or why not?
15. Can a single scientist observe all of the stages of secondary succession on an abandoned field? Explain your answer.

Concept Mapping

16. Use the following terms to create a concept map: abandoned farmland, lichens, bare rock, soil formation, horseweed, succession, forest fire, primary succession, secondary succession, pioneer species.

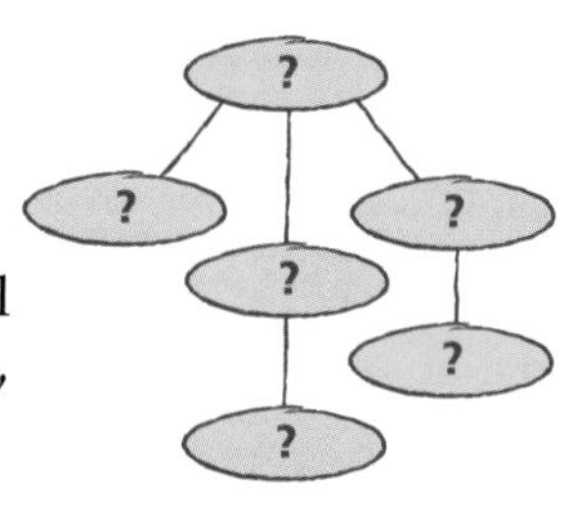

CRITICAL THINKING AND PROBLEM SOLVING

Write one or two sentences to answer the following questions:

17. Explain how living things would be affected if the water on our planet suddenly stopped evaporating.

18. How would living things be affected if there were no decomposers to cycle carbon back to the atmosphere?

19. Explain how living things would be affected if the bacteria responsible for nitrogen fixation were to die.

20. Describe why a lawn doesn't go through succession.

MATH IN SCIENCE

In 1996, 129 million metric tons of fertilizer were used world-wide. Use the following information to answer items 21, 22, and 23: 1996 world population = 5.7 billion; 1 metric ton = 1,000 kg; 1 kg = 2.2 lb.

21. Write out the number corresponding to 5.7 billion. How many zeros are in the number?

22. How many kilograms of fertilizer were used per person in 1996?

23. How many pounds of fertilizer were used per person?

INTERPRETING GRAPHICS

The following graph illustrates the concentration of carbon dioxide in the atmosphere from 1958 to 1994:

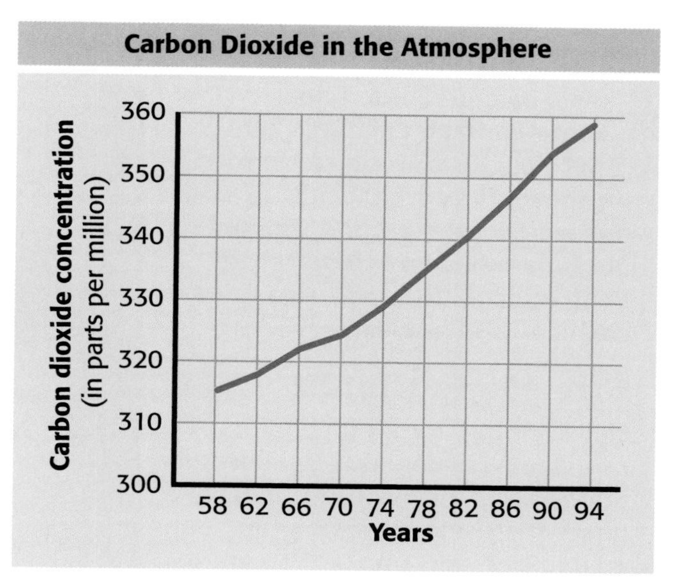

24. What was the concentration of carbon dioxide in parts per million in 1960? in 1994?

25. Is the concentration of carbon dioxide increasing or decreasing? Explain.

26. If the level of carbon dioxide continues to change at the same steady rate, what might be the concentration in 2010?

Reading Check-up

Take a minute to review your answers to the Pre-Reading Questions found at the bottom of page 28. Have your answers changed? If necessary, revise your answers based on what you have learned since you began this chapter.

WEIRD SCIENCE

WEATHER FROM FIRE

As a wildfire burned near Santa Barbara, California, in 1993, huge storm clouds formed overhead. Fiery whirlwinds danced over the ground. The fire not only was destroying everything in its path—it was also creating its own weather!

▲ *This towering whirlwind is lifting burning debris from a forest fire in Idaho.*

Fire-Made Clouds

Hot air rising from a forest fire can create tremendous updrafts. Surrounding air rushes in underneath the rising air, stirring up columns of ash, smoke, hot air, and noxious gases. Cool, dry air normally sinks down and stops these columns from developing any further. But if the conditions are just right, a surprising thing happens.

If the upper atmosphere contains warm, moist air, the moisture begins to condense on the ash and smoke. These droplets can develop into clouds. As the clouds grow, the droplets begin to collide and combine until they are heavy enough to fall as rain. The result is an isolated rainstorm, complete with thunder and lightning.

Whirlwinds of Fire

Forest fires can also create whirlwinds. These small, tornado-like funnels can be extremely dangerous. Whirlwinds are similar to dust devils that dance across desert sands. Their circular motion is created by an updraft that is forced to turn after striking an obstacle, such as a cliff or hill. Whirlwinds move across the ground at 8–11 km/h, sometimes growing up to 120 m high and 15 m wide.

Most whirlwinds last less than a minute, but they can cause some big problems. Firefighters caught in the path of whirlwinds have been severely injured and even killed. Also, if a whirlwind is hot enough, it can suck up tremendous amounts of air. The resulting updraft can pull burning debris up through the whirlwind. In some cases, burning trees have been uprooted and shot into the air. When the debris lands, it often starts new fires hundreds of meters away.

Think About It

► Fires are a natural part of the growth of a forest. For example, some tree seeds are released only under the extreme temperatures of a fire. Some scientists believe that forest fires should be allowed to run their natural course. Others argue that forest fires cause too much damage and should be extinguished as soon as possible. Do some additional research and then decide what you think.

EYE ON THE ENVIRONMENT

The Mysterious Dead Zone

Every summer, millions of fish are killed in an area in the Gulf of Mexico called a hypoxia region. Hypoxia is a condition that occurs when there is an unusually low level of oxygen in the water. The area is often referred to as the "dead zone" because almost every fish and crustacean in the area dies. In 1995, this zone covered more than 18,000 km^2, and almost 1 million fish were killed in a single week. Why does this happen? Can it be stopped?

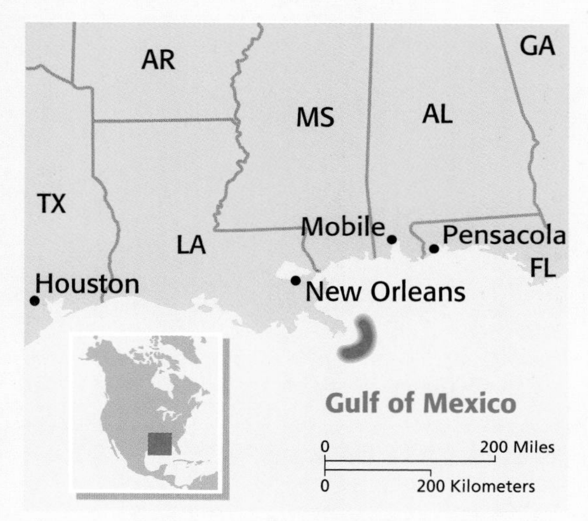

▲ *The Gulf of Mexico hypoxia region grew to the size of New Jersey in 1995.*

What's Going On?

When the oxygen levels in water drop drastically, the fish die. In the Gulf of Mexico hypoxia region, the water contains unusually high amounts of nitrogen and phosphorus. The nitrogen and phosphorus act as nutrients for the growth of algae. When the algae die, their bodies are decomposed by a large number of oxygen-consuming bacteria.

Scientists think the excess nitrogen and phosphate is from animal waste and runoff from farms and developments. The pollution may also be caused in part by the overfertilization of crops. The extra fertilizer runs into the rivers, which empty into the Gulf Coast.

Ecosystem Models Suggest Solutions

All along the Gulf Coast, marine scientists and Earth scientists are trying to find methods to reduce or eliminate the "dead zone." They have made physical and computer models of the Mississippi River ecosystem that have accurately predicted the data that has since been collected. The scientists have changed the models to see what happens. For example, wetlands are one of nature's best filters. They take up a lot of the chemicals present in water. Scientists predict that adding wetlands to the Mississippi River watershed could reduce the chemicals reaching the Gulf of Mexico. Although scientific models support this hypothesis, they also indicate that adding wetlands to the Mississippi River watershed would not be enough to completely prevent the "dead zone."

Find Some Solutions

► The Gulf of Mexico is not the only place that suffers from a hypoxia region. Research other bodies of water to find out how widespread the problem is. Have scientists found ways to reduce or eliminate the hypoxia regions elsewhere? How could this information be used to improve the situation in the Gulf of Mexico?

The Earth's Ecosystems

Sections

1. What are the main differences between a desert and a rain forest?
2. Where does the water in a lake come from?
3. Which has more species of plants and animals—the open ocean or a swamp? Why?

In Living Color

A flurry of orange fish swim through the sun-dappled crevices of a tropical coral reef. All around them other life exists—sea fans, eels, anemones, and living corals. Could this scene exist anywhere else? It could . . . if the place was underwater in a warm climate close to the shore. In this chapter, you will learn how the nonliving environment affects organisms and how they are adapted to where they live.

A MINI-ECOSYSTEM

In this activity, you will build and observe a miniature ecosystem.

Procedure

1. Place a layer of **gravel** in the bottom of a **large widemouth jar** or **2 L bottle** with the top cut off. Add a layer of **soil.**
2. Add a variety of **small plants** that require similar growing conditions. Choose plants that will not grow too quickly.
3. Spray **water** inside the jar to moisten the soil.
4. Cover the jar, and place it in indirect light. Describe the appearance of your ecosystem in your ScienceLog.
5. Observe your mini-ecosystem every week. Spray it with water to keep the soil moist. Record all of your observations.

Analysis

6. List all of the nonliving factors in the ecosystem you have created.
7. How is your mini-ecosystem similar to a real ecosystem? How is it different?

SECTION 1
READING WARM-UP

Terms to Learn

abiotic
biome
savanna
desert
tundra
permafrost

What You'll Do

- Define *biome.*
- Describe three different forest biomes.
- Distinguish between temperate grasslands and savannas.
- Describe the importance of permafrost to the arctic tundra biome.

Land Ecosystems

Imagine that you are planning a camping trip. You go to a travel agency, where you find a virtual-reality machine that can let you experience different places before you go. You put on the virtual-reality gear, and suddenly you are transported. At first your eyes hurt from the bright sunlight. The wind that hits your face is very hot and very dry. As your eyes grow accustomed to the light, you see a large cactus to your right and some small, bushy plants in the distance. A startled jack rabbit runs across the dry, dusty ground. A lizard basks on a rock. Where are you?

You may not be able to pinpoint your exact location, but you probably realize that you are in a desert. That's because most deserts are hot and dry. These **abiotic,** or nonliving, factors influence the types of plants and animals that live in the area.

The Earth's Biomes

A desert is one of Earth's biomes. A **biome** is a geographic area characterized by certain types of plant and animal communities. A biome contains a number of smaller but related ecosystems. For example, a tropical rain forest is a biome that contains river ecosystems, treetop ecosystems, forest-floor ecosystems, and many others. A biome is not a specific place. For example, a desert biome does not refer to a particular desert. A desert biome refers to any and all desert ecosystems on Earth. The major biomes of Earth are shown in **Figure 1.**

Figure 1 *Rainfall and temperature are the main factors that determine what biome is found in a region. What kind of biome do you live in?*

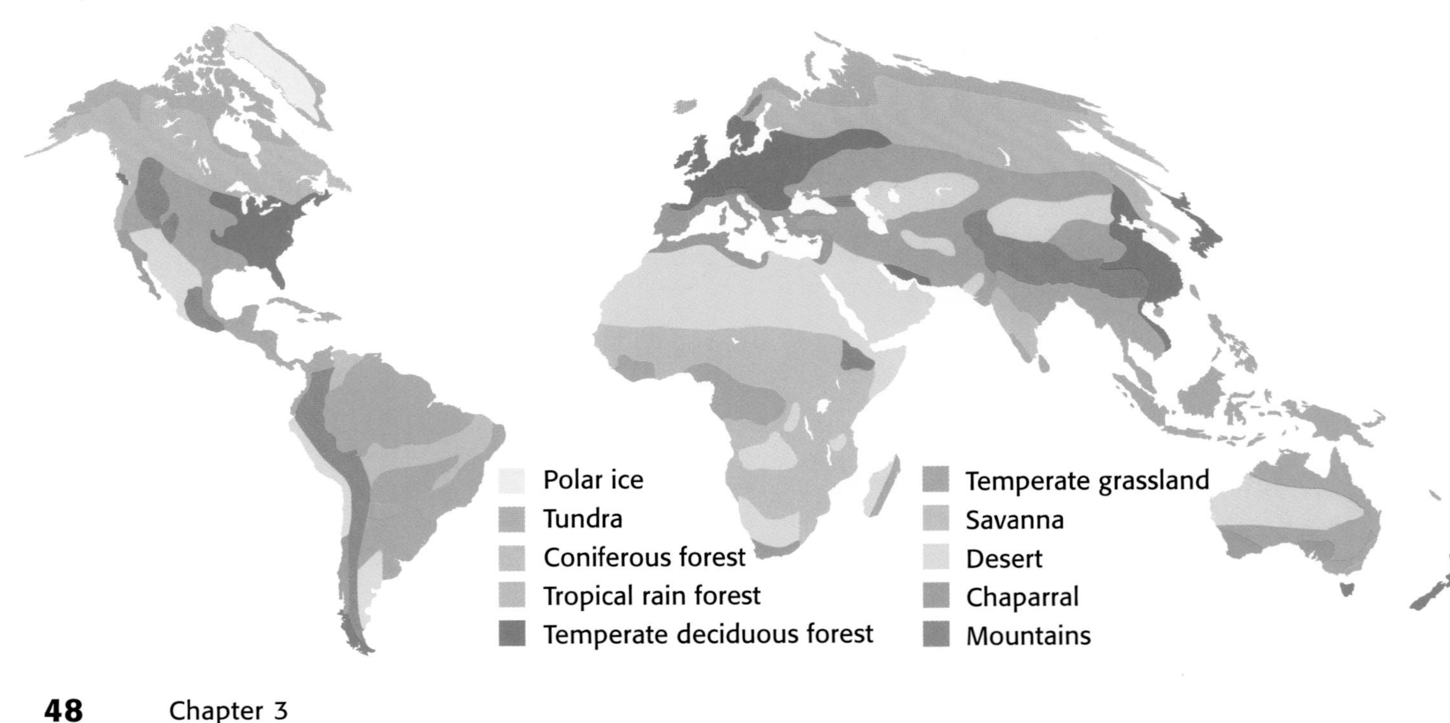

Forests

Forest biomes develop where there is enough rain and where the temperature is not too hot in the summer or too cold in the winter. There are three main types of forest biomes—temperate deciduous forests, coniferous forests, and tropical rain forests. The type of forest that develops depends on the area's temperature and rainfall.

Temperate Deciduous Forest

Average Yearly Rainfall
75–125 cm (29.5–49 in.)

Average Temperatures
Summer: 28°C (82.4°F)
Winter: 6°C (42.8°F)

Temperate Deciduous Forests In the autumn, have you seen leaves that change colors and fall from trees? If so, you have seen trees that are *deciduous,* which comes from a Latin word meaning "to fall off." By losing their leaves in the fall, deciduous trees are able to conserve water during the winter. **Figure 2** shows a temperate deciduous forest. Most of these forests contain several different species of trees. Temperate deciduous forests also support a variety of animals, such as bears and woodpeckers.

Figure 2 *In a temperate deciduous forest, mammals, birds, and reptiles thrive on the abundance of leaves, seeds, nuts, and insects.*

In forests, plant growth occurs in layers. The leafy tops of the trees reach high above the forest floor, where they receive full sunlight.

Beneath the tree layer, woody shrubs and bushes catch the light that filters through the trees.

Grasses, herbs, ferns, and mosses are scattered across the forest floor. Most of the flowering plants bloom, and produce seeds in early spring, before the trees grow new leaves.

Coniferous Forest

Average Yearly Rainfall
35–75 cm (14–29.5 in.)

Average Temperatures
Summer: 14°C (57.2°F)
Winter: −10°C (14°F)

Coniferous Forests Coniferous forests do not change very much from summer to winter. They are found in areas with long, cold winters. These forests consist mainly of *evergreen* trees, which are trees that don't lose their leaves and stay green all year. Most of these trees are *conifers,* which means that they produce seeds in cones. You have probably seen a pine cone. Pine trees are common conifers.

Most conifers can also be identified by their compact, needlelike leaves. These leaves, or needles, have a thick waxy coating that prevents them from drying out and being damaged during winter.

Figure 3 shows a coniferous forest and some of the animals that live there. Notice that not many large plants grow beneath the conifers, partly because very little light reaches the ground.

Figure 3 *Many animals that live in a coniferous forest survive the harsh winters by hibernating or migrating to a warmer climate for the winter.*

Tropical Rain Forests The tropical rain forest has more biological *diversity* than any other biome on the planet; that is, it contains more species than any other biome. As many as 100 species of trees may live in an area about one-fourth the size of a football field. Although some animals live on the ground, the treetops, or *canopy*, are the preferred living site. A huge variety of animals live in the canopy. If you counted the birds in the canopy of a rain forest, you would find up to 1,400 species! **Figure 4** shows some of the diversity of the tropical rain forest biome.

Most of the nutrients in a tropical rain forest biome are in the vegetation. The topsoil is actually very thin and poor in nutrients. Farmers who cut down the forest to grow crops must move their crops to freshly cleared land after about 2 years.

Tropical Rain Forest

Average Yearly Rainfall
Up to 400 cm (157.5 in.)

Average Temperatures
Daytime: 34°C (93°F)
Nighttime: 20°C (68°F)

Figure 4 A Tropical Rain Forest Biome

Grasslands

Plains, steppes, savannas, prairies, pampas—these are names for regions where grasses are the major type of vegetation. Grasslands are found between forests and deserts. They exist on every continent. Most grasslands are flat or have gently rolling hills.

Temperate Grassland

Average Yearly Rainfall
25–75 cm (10–29.5 in.)

Average Temperatures
Summer: 30°C (86°F)
Winter: 0°C (32°F)

Temperate Grasslands Temperate grassland vegetation is mainly grasses mixed with a variety of flowering plants. There are few trees because fires prevent the growth of most slow-growing plants. The world's temperate grasslands support small, seed-eating mammals, such as prairie dogs and mice, and large herbivores, such as the bison of North America, shown in **Figure 5.**

Figure 5 *Bison roamed the temperate grasslands in great herds before they were hunted nearly to extinction.*

Savanna

Average Yearly Rainfall
150 cm (59 in.)

Average Temperatures
Dry season: 34°C (93°F)
Wet season: 16°C (61°F)

Savanna The **savanna** is a tropical grassland with scattered clumps of trees. During the dry season, the grasses die back, but the deep roots survive even through months of drought. During the wet season, the savanna may receive as much as 150 cm of rain. The savannas of Africa are inhabited by the most abundant and diverse groups of large herbivores in the world, like those shown in **Figure 6.** These include elephants, giraffes, zebras, gazelles, and wildebeests.

Figure 6 *Carnivores, such as lions and leopards, prey on herbivores, such as these zebras and wildebeests. Hyenas and vultures usually "clean up" after the carnivores.*

Self-Check

Use the map in Figure 1 to compare the locations of deciduous and coniferous forests. Explain the differences in location between the two biomes. *(See page 168 to check your answers.)*

How do animals survive in the heat of the desert? Quite nicely, thank you! See how on page 132 of your LabBook.

Deserts

Deserts are hot, dry regions that support a variety of plants and animals. In a desert, most of the water that falls to the ground evaporates. Organisms have evolved in specialized ways to survive extreme temperatures with very little water. For example, plants grow far apart to reduce competition for the limited water supply. Some plants have shallow, widespread roots that absorb water quickly during a storm, while others may have very deep roots that reach ground water.

Animals also have adaptations for survival in the desert. Most are active only at night, when temperatures are cooler. Tortoises eat the flowers or leaves of plants and store the water under their shells for months. **Figure 7** shows how some desert plants and animals survive in the heat with little water.

Desert

Average Yearly Rainfall
Less than 25 cm (10 in.)

Average Temperatures
Summer: 38°C (100°F)
Winter: 7°C (45°F)

Figure 7 *There are many well-adapted residents of the desert biome.*

Tundra

Tundra

Average Yearly Rainfall
30–50 cm (12–20 in.)

Average Temperatures
Summer: 12°C (53.6°F)
Winter: −26°C (−14°F)

In the far north and on the tops of high mountains, the climate is so cold that no trees can grow. A biome called the **tundra** is found there.

Arctic Tundra The major feature of the arctic tundra is permafrost. During the short growing season, only the surface of the soil thaws. The soil below the surface, the **permafrost,** stays frozen all the time. Even though there is little rainfall, water is not in short supply. That's because the permafrost prevents the rain that does fall from draining, and the surface soil stays wet and soggy. Lakes and ponds are common.

The layer of unfrozen soil above the permafrost is too shallow for deep-rooted plants to survive. Grasses, sedges, rushes, and small woody shrubs are common. A layer of mosses and lichens grows beneath these plants on the surface of the ground. Tundra animals, like the one shown in **Figure 8,** include large mammals such as caribous, musk oxen, and wolves, as well as smaller animals, such as lemmings, shrews, and hares. Migratory birds are abundant in summer.

Figure 8 *Caribou migrate to more plentiful grazing grounds during long, cold winters in the tundra.*

Alpine Tundra Another tundra biome is found above the tree line of very high mountains. These areas, called alpine tundra, receive a lot of sunlight and precipitation, mostly in the form of snow.

MATH BREAK

Rainfall

In 1 year, what is the difference in the rainfall amounts in a coniferous forest, a tropical rain forest, a desert, and a savanna? To compare, create a bar graph of the rainfall in each biome from the data given in this section.

SECTION REVIEW

1. How is the climate of temperate grasslands different from that of savannas?
2. Describe three ways that plants and animals are adapted to the desert climate.
3. Where are most of the nutrients in a tropical rain forest?
4. **Applying Concepts** Could arctic tundra accurately be called a frozen desert? Why or why not?

READING WARM-UP

Marine Ecosystems

Terms to Learn

marine
phytoplankton
zooplankton
estuary

What You'll Do

- Distinguish between the different areas of the ocean.
- Explain the importance of plankton in marine ecosystems.
- Describe coral reefs and intertidal areas.

They cover almost three-quarters of Earth's surface and contain almost 97 percent of Earth's water supply. The largest animals on Earth inhabit them, along with billions of microscopic creatures, shown in **Figure 9.** Their habitats range from dark, cold, high-pressure depths to warm sandy beaches; from icy polar waters to rocky coastlines. They are oceans and seas. Wherever these salty waters are found, marine ecosystems are found. A **marine** ecosystem is one that is based on salty water. This abiotic factor has a strong influence on the ecosystems of oceans and seas.

Abiotic Factors Rule

Like terrestrial biomes, marine biomes are shaped by abiotic factors. These include temperature, the amount of sunlight penetrating the water, the distance from land, and the depth of the water. These abiotic factors are used to define certain areas of the ocean. As with terrestrial biomes, marine biomes occur all over Earth and can contain many ecosystems.

Sunny Waters Water absorbs light, so sunlight can penetrate only about 200 m below the ocean's surface, even in the clearest water. As you know, most producers use photosynthesis to make their own food. Because photosynthesis requires light, most producers are found only where light penetrates. The most abundant producers in the ocean are called **phytoplankton.** Phytoplankton are microscopic photosynthetic organisms that float near the surface of the water. Using the energy of sunlight, these organisms make their own food just as plants that live on land do. **Zooplankton** are the consumers that feed on the phytoplankton. They are small animals that, along with phytoplankton, form the base of the oceans' feeding relationships.

Figure 9 *Marine ecosystems support a broad diversity of life, from the humpback whale to microscopic phytoplankton.*

Wonderful Watery Biomes

Unique and beautiful biomes exist in every part of oceans and seas. These biomes are home to many unusually adapted organisms. The major ocean areas and some of the organisms that live in them are shown below in **Figure 10.**

A The Intertidal Zone The intertidal zone is the area where the ocean meets the land. This area is above water part of the day, when the tide is out, and is often battered by waves. Mud flats, rocky shores, and sandy beaches are all in the intertidal area.

B The Neritic Zone Moving seaward, the water becomes gradually deeper toward the edge of the continental shelf. Water in this area is generally less than 200 m deep and usually receives a lot of sunlight. Diverse and colorful coral reefs exist in the waters over the continental shelf, where the water is warm, clear, and sunny.

Figure 10 *The life in a particular area depends on how much light the area receives, how far the area is from land, and how far the area is beneath the surface.*

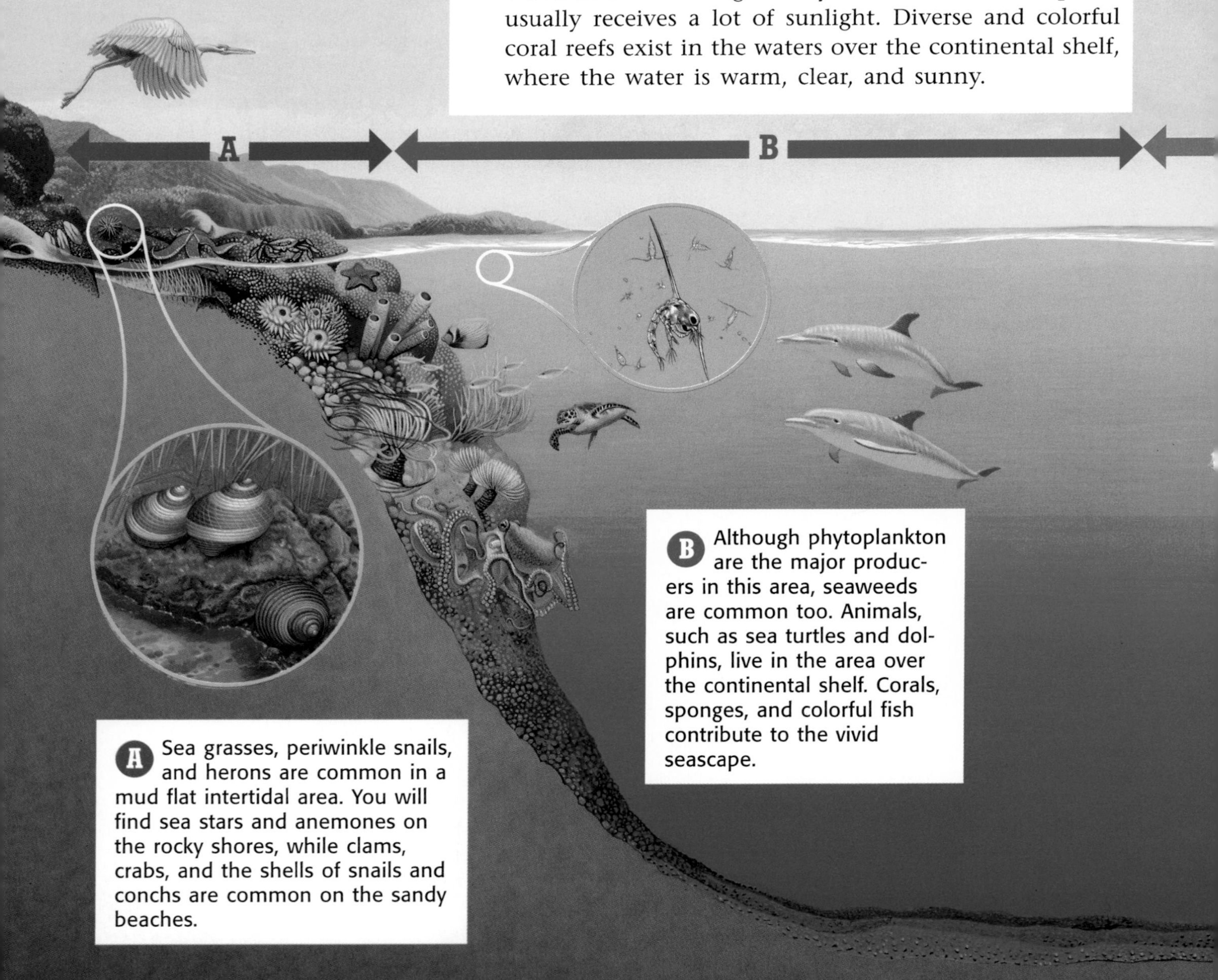

A

B

A Sea grasses, periwinkle snails, and herons are common in a mud flat intertidal area. You will find sea stars and anemones on the rocky shores, while clams, crabs, and the shells of snails and conchs are common on the sandy beaches.

B Although phytoplankton are the major producers in this area, seaweeds are common too. Animals, such as sea turtles and dolphins, live in the area over the continental shelf. Corals, sponges, and colorful fish contribute to the vivid seascape.

C The Oceanic Zone Past the continental shelf, the sea floor drops sharply. This is the deep water of the open ocean. To a depth of about 200 m, phytoplankton are the producers. At greater depths, no light penetrates, so most organisms obtain energy by consuming organic material that falls from the surface.

D The Benthic Zone The benthic zone is the sea floor. It extends from the upper edge of the intertidal zone to the bottom of the deepest ocean waters. Organisms that live on the deep-sea floor obtain food mostly by consuming material that filters from above. Some bacteria are *chemosynthetic,* which means they use chemicals in the water near thermal vents to make food. A thermal vent is a place on the ocean floor where heat escapes through a crack in the Earth's crust.

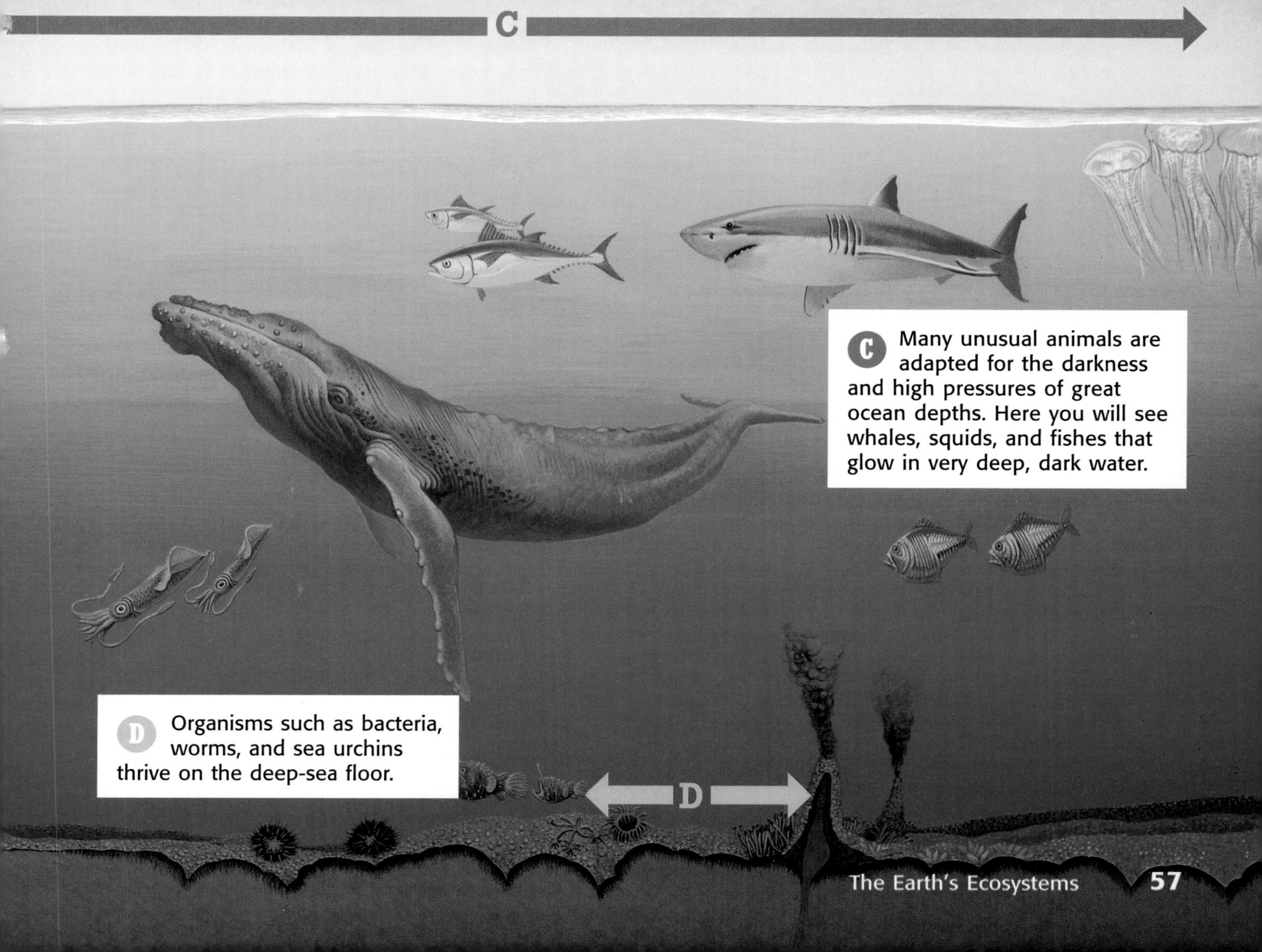

C

C Many unusual animals are adapted for the darkness and high pressures of great ocean depths. Here you will see whales, squids, and fishes that glow in very deep, dark water.

D Organisms such as bacteria, worms, and sea urchins thrive on the deep-sea floor.

D

Figure 11 *A coral reef is one of the most biologically diverse biomes.*

A Closer Look

Marine environments provide most of the water for Earth's rainfall through evaporation and precipitation. Ocean temperatures and currents have major effects on world climates and wind patterns. Humans harvest enormous amounts of food from the oceans and dump enormous amounts of waste into them. Let's take a closer look at some of the special environments that thrive in the ocean.

Coral Reefs In some sunny tropical waters, the sea floor contains coral reefs. Corals live in a close relationship with single-celled algae. The algae produce organic nutrients through photosynthesis. This provides food for the coral. The coral provide a place in the sun for the algae to live. The foundation of the reef is formed from coral skeletons that have built up over thousands of years. Coral reefs, like the one in **Figure 11,** are home to many marine species, including a large variety of brightly colored fish and organisms such as sponges and sea urchins.

Figure 12 *The Sargasso Sea is a spawning place for eels and home to a rich diversity of organisms.*

The Sargasso Sea In the middle of the Atlantic Ocean is a large ecosystem with no land boundaries. It is called the Sargasso Sea. *Sargassum* is a type of algae usually found attached to rocks on the shores of North America, but it forms huge floating rafts in the Sargasso Sea. Animals adapted to this environment live among the algae. Most of the animals are the same color as the *Sargassum*. Some even look like it! Why do you think this is so? Can you find a fish in **Figure 12**?

Self-Check

1. List three factors that characterize marine biomes.
2. Describe one way organisms obtain energy at great depths in the open ocean.

(See page 168 to check your answers.)

Polar Ice The Arctic Ocean and the open waters surrounding Antarctica make up a very unusual marine biome—one that includes ice!

The icy waters are rich in nutrients from the surrounding landmasses. These nutrients support large populations of plankton. The plankton in turn support a great diversity of fish, birds, and mammals, as shown in **Figure 13.**

Figure 13 *Sea lions and penguins are some of the animals found on the shores of Antarctica.*

Estuaries An area where fresh water from streams and rivers spills into the ocean is called an **estuary.** The fresh water constantly mixes with the salt water of the sea. The amount of salt in an estuary changes frequently. When the tide rises, the salt content of the water rises. When the tide recedes, the water becomes fresher. The fresh water that spills into an estuary is rich in nutrients that are carried by water running off the land. Because estuaries are so nutrient-rich, they support large numbers of plankton, which provide food for many larger animals.

Intertidal Areas Intertidal areas include mudflats, sandy beaches, and rocky shores. Mud flats are home to many worms and crabs and the shorebirds that feed on them. Sandy beaches are also home to worms, clams, crabs, and plankton that live among the sand grains.

On rocky shores, organisms either have tough holdfasts or are able to cement themselves to a rock to avoid being swept away by crashing waves. **Figure 14** shows some of these animals.

Figure 14 *Sea stars can wedge themselves under a rock to keep from being washed out to sea.*

SECTION REVIEW

1. Explain how a coral reef is both living and dead.
2. Why do estuaries support such an abundance of life?
3. **Analyzing Relationships** Explain how the amount of light an area receives determines the kinds of organisms that live in the open ocean.

Freshwater Ecosystems

Terms to Learn

tributary
littoral zone
open-water zone
deep-water zone
wetland
marsh
swamp

What You'll Do

- List the characteristics of rivers and streams.
- Describe the littoral zone of a pond.
- Distinguish between two types of wetlands.

A mountain brook bubbles over rocks down a mountainside. A mighty river thunders through a canyon. A small pond teems with life. A lake tosses boats during a heavy storm. A dense swamp echoes with the sounds of frogs and birds.

What do all of these places have in common? They are freshwater ecosystems. Like other ecosystems, freshwater ecosystems are characterized by abiotic factors, primarily the speed at which the water is moving.

Water on the Move

Brooks, streams, and rivers are ecosystems based on moving water. The water may begin flowing from melting ice or snow. Or it may come from a spring, where water flows up to the surface of the Earth. Each trickle or stream of water that joins a larger trickle or stream is a **tributary.**

Fast-Moving Water As more tributaries join a stream, the stream becomes larger and wider, forming a river. Aquatic plants line the edge of the river. Fishes live in the open waters. In the mud at the bottom, burrowers, such as freshwater clams and mussels, make their home.

Organisms that live in moving water require special adaptations to avoid being swept away with the current. Producers, such as algae and moss, cling to rocks. Consumers, such as insect larvae, live under rocks in the shallow water. Some consumers, such as tadpoles, use suction disks to hold themselves to rocks.

Slowing Down As a river grows wider and slower, it may *meander* back and forth across the landscape. Organic material and sediment may be deposited on the bottom, building *deltas*. Dragonflies, water striders, and other invertebrates live in and on slow-moving water. Eventually, the moving water empties into a lake or an ocean. **Figure 15** shows how a river can grow from melted snow.

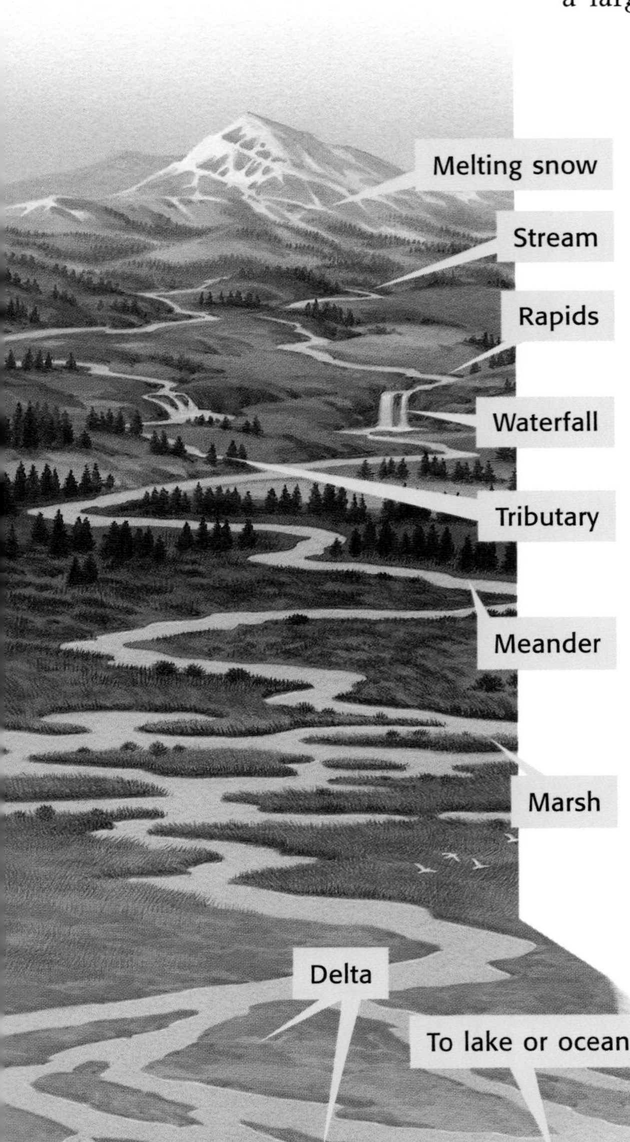

Figure 15 *This figure shows the features of a typical river. Where is the water moving rapidly? Where is it moving slowly?*

Still Waters

Ponds and lakes have different ecosystems than streams and rivers have. Lake Superior, the largest lake in the world, has more in common with a small beaver pond than with a river. **Figure 16** shows a cross section of a typical lake. In looking at this illustration, you will notice that the lake has been divided into three zones. As you read on, you will learn about these zones and the ecosystems they contain.

Pond Food Connections

1. On **index cards**, write the names of the animals and plants that live in a typical freshwater pond or small lake. Write one type of organism on each card.
2. Use **yarn** or **string** to connect each organism to its food sources.
3. In your ScienceLog, describe the food relationships in the pond.

Where Water Meets Land Look at Figure 16 again, and locate the **littoral zone.** It is the zone closest to the edge of the land. This zone has many inhabitants. Plants that grow in the water closest to the shore include cattails and rushes. Farther from the shore are floating leaf plants, such as water lilies. Still farther out are submerged pond weeds that grow beneath the surface of the water.

The plants of the littoral zone provide a home for small animals, such as snails, small arthropods, and insect larvae. Clams, worms, and other organisms burrow in the mud. Frogs, salamanders, water turtles, various kinds of fishes, and water snakes also live in this area.

Life at the Top Look again at Figure 16. This time locate the **open-water zone.** This zone extends from the littoral zone across the top of the water. The open-water zone only goes as deep as light can reach. This is the habitat of bass, blue gills, lake trout, and other fish. Phytoplankton are the most abundant photosynthetic organisms in the open-water zone of a lake.

Life at the Bottom Now look at Figure 16 and find the **deep-water zone.** This zone is below the open-water zone, where no light reaches. Catfish, carp, worms, insect larvae, crustaceans, fungi, and bacteria live here. These organisms feed on dead organic material that falls down from above.

Figure 16 *Freshwater ecosystems are characterized by abiotic factors that determine which organisms live there.*

A Trip to Lake Superior

Suppose you are a life scientist who specializes in the plants that live in and near Lake Superior. You are preparing for a yearlong expedition to Thunder Bay, on the Canadian shore of Lake Superior. You will stay "in the wild." Based on what you have learned about ecosystems, answer the following questions: How will you live while you are there? What will you bring along? What problems will you encounter? How will you overcome them?

Wetlands

A **wetland** is an area of land where the water level is near or above the surface of the ground for most of the year. Wetlands support a variety of plant and animal life. They also play an important role in flood control. During heavy rains or spring snow melt, wetlands soak up large amounts of water. The water in wetlands also seeps into the ground, replenishing underground water supplies.

Activity

While exploring in a wetland, you have discovered a new organism. In your ScienceLog, draw the organism. Describe what it looks like and how it is adapted to its environment. Trade with a partner. Is your partner's organism believable?

TRY at HOME

Marshes A **marsh** is a treeless wetland ecosystem where plants such as cattails and rushes grow. A freshwater marsh is shown in **Figure 17.** Freshwater marshes are found in shallow waters along the shores of lakes, ponds, rivers, and streams. The plants in a marsh vary depending on the depth of the water and the location of the marsh. Grasses, reeds, bulrushes, and wild rice are common marsh plants. Muskrats, turtles, frogs, and red-wing blackbirds can be found living in marshes.

Figure 17 *Turtles find a lot of places to escape from predators in a freshwater marsh. Many species raise their young in these protected areas.*

Swamps A **swamp** is a wetland ecosystem where trees and vines grow. Swamps occur in low-lying areas and beside slow-moving rivers. Most swamps are flooded only part of the year, depending on the rainfall. Trees may include willows, bald cypresses, water tupelos, oaks, and elms. Vines such as poison ivy grow up trees, and Spanish moss hangs from the branches. Water lilies and other lake plants may grow in open-water areas. Swamps, like the one in **Figure 18,** provide a home for a variety of fish, snakes, and birds.

Figure 18 *The bases of the trunks of these trees are adapted to give the tree more support in the wet, soft sediment under the water in this swamp.*

From Lake to Forest

How can a lake or pond, like the one in **Figure 19,** disappear? Water entering a standing body of water usually carries nutrients and sediment along with it. These materials then settle to the bottom. Dead leaves from overhanging trees and decaying plant and animal life also settle to the bottom. Gradually, the pond or lake fills in. Plants grow in the newly filled areas, closer and closer toward the center. With time, the standing body of water becomes a marsh. Eventually, the marsh turns into a forest.

Figure 19 *Eventually decaying organic matter, along with sediment in the runoff from land, will fill in this pond.*

SECTION REVIEW

1. Describe some adaptations of organisms that live in moving water.
2. Compare the littoral zone with the open-water zone of a pond.
3. How is a swamp different from a marsh?
4. **Analyzing Concepts** The center of a pond is 10 m deep. Near the shore it is 0–1 m deep. Describe the types of organisms that might live in each zone.

internet**connect**

SCILINKS NSTA

TOPIC: Freshwater Ecosystems
GO TO: www.scilinks.org
***sci*LINKS NUMBER:** HSTL495

Skill Builder Lab

Too Much of a Good Thing?

Plants require nutrients, such as phosphates and nitrates. Phosphates are often found in detergents. Nitrates are often found in animal wastes and fertilizers. When large amounts of these nutrients enter rivers and lakes, algae and plant life grow rapidly and then die off. Microorganisms that decompose the dead matter use up oxygen in the water, killing fish and other animals. In this activity, you will observe the effect of fertilizers on organisms that live in pond water.

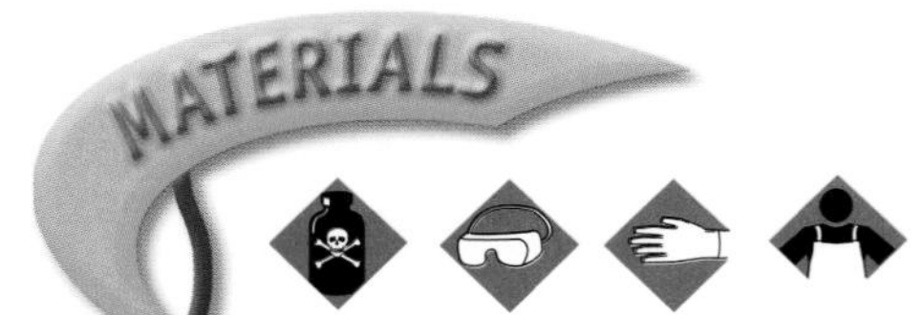

- wax pencil
- 1 qt (or 1 L) jars (3)
- 2.25 L of distilled water
- fertilizer
- graduated cylinder
- stirring rod
- 300 mL of pond water containing living organisms
- eyedropper
- microscope
- microscope slides with coverslips
- plastic wrap
- protective gloves

Procedure

1. Use a wax pencil to label one jar "Control," the second jar "Fertilizer," and the third jar "Excess Fertilizer."

2. Pour 750 mL of distilled water in each of the jars. Read the label on the fertilizer container to determine the recommended amount of fertilizer. To the Fertilizer jar, add the amount of fertilizer recommended for 750 mL of water. To the Excess Fertilizer jar, add 10 times the recommended amount. Stir the contents of each jar to dissolve the fertilizer.

3. Obtain a sample of pond water. Stir it gently but thoroughly to make sure that the organisms in it are evenly distributed. Pour 100 mL of pond water into each of the three jars.

4. Observe a drop of pond water from each jar under the microscope. Draw at least four of the organisms. Determine whether the organisms you see are algae, which are usually green, or consumers, which are usually able to move. Describe the number and type of organisms in the pond water.

Common Pond-Water Organisms

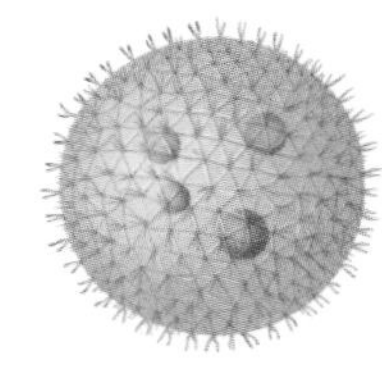
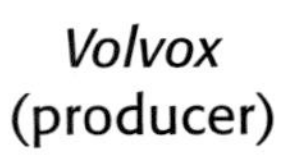
Volvox (producer)

Spirogyra (producer)

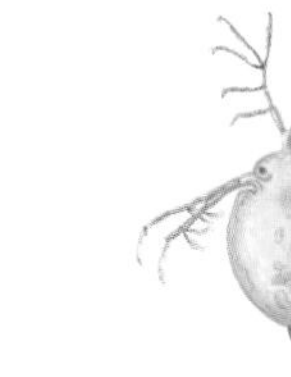
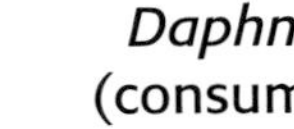
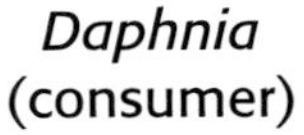
Daphnia (consumer)

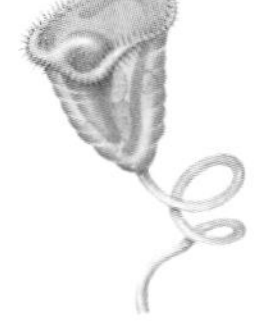
Vorticella (consumer)

5. Cover each jar loosely with plastic wrap. Place the jars near a sunny window, but do not place them in direct sunlight.

6. Based on your understanding of how ponds and lakes eventually fill up to become dry land, make a prediction about how the pond organisms will grow in each of the three jars.

7. Make three data tables in your ScienceLog. Be sure to allow enough space to record your observations. Title one table "Control," as shown below. Title another table "Fertilizer," and title the third table "Excess Fertilizer."

8. Observe the jars when you first set them up and at least once every three days for the next three weeks. Note the color, odor, and any visible presence of organisms. Record your observations.

9. When organisms begin to be visible in the jars, use an eyedropper to remove a sample from each jar, and observe the sample under the microscope. How have the number and type of organisms changed since you first looked at the pond water? Record your observations.

10. At the end of the three-week period, remove a sample from each jar and observe each sample under the microscope. Draw at least four of the most abundant organisms, and describe how the number and type of organisms have changed since your last microscopic observation.

Analysis

11. After three weeks, which jar has the most abundant growth of algae? What may have caused this growth?

12. Did you observe any effects on organisms (other than the algae) in the jar with the most abundant algal growth? Explain your answer.

13. Did your observations match your prediction? Explain your answer.

14. How might the rapid filling of natural ponds and lakes be prevented or slowed?

Control			
Date	Color	Odor	Other observations

Chapter Highlights

SECTION 1

Vocabulary

abiotic *(p. 48)*
biome *(p. 48)*
savanna *(p. 52)*
desert *(p. 53)*
tundra *(p. 54)*
permafrost *(p. 54)*

Section Notes

- Rainfall and temperature are the main factors that determine what kind of biome is found in a region.
- The three main forest biomes are the temperate deciduous forest and the coniferous forest, which experience warm summers and cold winters, and the tropical rain forest, where temperatures stay warm.
- Grasslands receive more rain than deserts and receive less rain than forests. Temperate grasslands have hot summers and cold winters. Savannas have wet and dry seasons.
- Deserts receive less than 25 cm of rain a year. Plants and animals competing for the limited water supply have developed special adaptations for survival.
- The tundra biome is found mainly in the Arctic region. Arctic tundra is characterized by permafrost.

Labs

Life in the Desert *(p. 132)*

SECTION 2

Vocabulary

marine *(p. 55)*
phytoplankton *(p. 55)*
zooplankton *(p. 55)*
estuary *(p. 59)*

Section Notes

- The kinds of marine organisms that inhabit an area vary depending on the water depth, the temperature, the amount of light, and the distance from shore.
- The intertidal area is the area where sea and land meet.
- The sea floor is home to biomes as different as coral reefs and thermal vents.
- The open ocean includes unique biomes, including the Sargasso Sea and the cold water oceans around the poles.

☑ Skills Check

Math Concepts

RAINFALL Using a meterstick, measure 400 cm on the floor of your classroom. This distance represents the depth of rainfall a rain forest receives per year. Next measure 25 cm. This measurement represents the amount of rainfall a desert receives per year. Compare these two quantities. Express your comparison as a ratio.

$$\frac{25}{400} = \frac{1}{16}$$

In 1 year, a desert receives $\frac{1}{16}$ the rainfall that a rain forest receives.

Visual Understanding

RAIN FOREST Look at Figure 4, on page 51. There are three layers of a rain forest—the upper story, the middle story, and the ground story. The upper story is the canopy, where most rain forest species live and where there is the most sunlight. The middle story is under the canopy and above the ground. The ground story is dark in most parts of the forest. Most plants in the rain forest grow very tall to compete for light in the canopy. Growth of plants on the ground story is not very dense due to the lack of available light.

SECTION 2

- An estuary is a region where fresh water from rivers spills into the ocean and the fresh and salt water mix with the rising and falling of the tides.

Labs

Discovering Mini-Ecosystems *(p. 133)*

SECTION 3

Vocabulary

tributary *(p. 60)*
littoral zone *(p. 61)*
open-water zone *(p. 61)*
deep-water zone *(p. 61)*
wetland *(p. 62)*
marsh *(p. 62)*
swamp *(p. 63)*

Section Notes

- Freshwater ecosystems are classified according to whether they have running water or standing water. Brooks, rivers, and streams contain running water. Lakes and ponds contain standing water.
- As tributaries join a stream between its source and the ocean, the volume of water in the stream increases, the nutrient content increases, and the speed decreases.
- The types of organisms found in a stream or river are determined mainly by how quickly the current is moving.
- The littoral zone of a lake is inhabited by floating plants. These plants provide a home for a rich diversity of animal life.
- Wetlands include marshes, which are treeless, and swamps, where trees and vines grow.

internetconnect

GO TO: go.hrw.com

Visit the **HRW** Web site for a variety of learning tools related to this chapter. Just type in the keyword:

KEYWORD: HSTECO

GO TO: www.scilinks.org

Visit the **National Science Teachers Association** on-line Web site for Internet resources related to this chapter. Just type in the ***sci*LINKS** number for more information about the topic:

TOPIC: Forests — ***sci*LINKS NUMBER:** HSTL480
TOPIC: Grasslands — ***sci*LINKS NUMBER:** HSTL485
TOPIC: Marine Ecosystems — ***sci*LINKS NUMBER:** HSTL490
TOPIC: Freshwater Ecosystems — ***sci*LINKS NUMBER:** HSTL495

Chapter Review

USING VOCABULARY

To complete the following sentences, choose the correct term from each pair of terms listed below:

1. At the edge of the __?__, the open ocean begins. *(continental shelf* or *Sargasso Sea)*

2. __?__ are tiny consumers that live in water. *(Phytoplankton* or *Zooplankton)*

3. A __?__ is a treeless wetland. *(swamp* or *marsh)*

4. __?__ lose their leaves in order to conserve water. *(Deciduous trees* or *Conifers)*

5. The major feature of the __?__ biome is permafrost. *(desert* or *tundra)*

6. Each major type of plant community and its associated animal communities make up a(n)__?__. *(estuary* or *biome)*

UNDERSTANDING CONCEPTS

Multiple Choice

7. The most numerous organisms in the oceans are the
   a. plankton.
   b. *Sargassum.*
   c. coral animals.
   d. marine mammals.

8. Marine ecosystems at the poles are unusual because
   a. animals spend time both in and out of the water.
   b. plankton are rare.
   c. they contain ice.
   d. the salt content of the water is very high.

9. The major factor that determines the types of organisms that live in a stream or river is
   a. the water temperature.
   b. the speed of the current.
   c. the depth of the water.
   d. the width of the stream or river.

10. Marine ecosystems
    a. contain the largest animals in the world.
    b. exist in all ocean zones.
    c. include environments where organisms survive without light.
    d. All of the above

11. Two major factors that determine what kind of a biome is found in a region are
    a. the amount of rainfall and the temperature.
    b. the depth of water and the distance from land.
    c. the wave action and the salt content of the water.
    d. All of the above

Short Answer

12. Describe how a stream changes as it moves from its source toward the ocean.

13. Describe two adaptations of animals to the desert environment.

14. Are wetlands always wet? Explain.

15. Explain how the salt content in an estuary changes constantly.

Concept Mapping

16. Use the following terms to create a concept map: tropical rain forest, deep-rooted plants, coral reef, canopy, biomes, permafrost, desert, continental shelf, tundra, ecosystems.

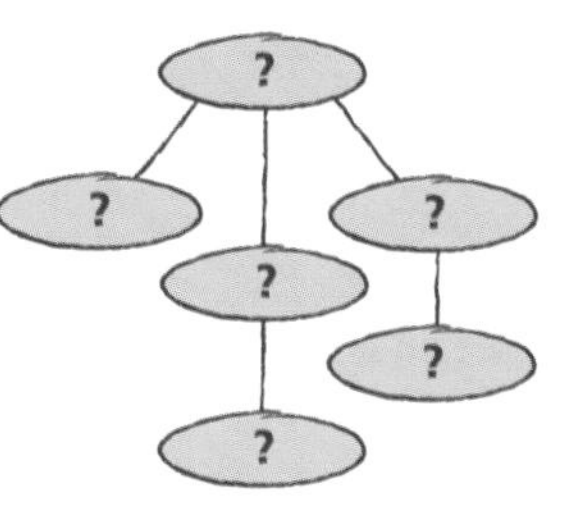

CRITICAL THINKING AND PROBLEM SOLVING

Write one or two sentences to answer the following questions:

17. While excavating a region now covered by grasslands, paleontologists discover the fossil remains of ancient fish and shellfish. What might they conclude?

18. In order to build a new shopping center, developers fill in a wetland. Afterward, flooding becomes a problem in this area. How can this be explained?

19. Explain why most desert flowering plants bloom, bear seeds, and die within a few weeks, while some tropical flowering plants remain in bloom for a much longer time.

MATH IN SCIENCE

20. What is the average difference in rainfall between a temperate deciduous forest and a coniferous forest?

21. An area of Brazilian rain forest received 347 cm of rain in one year. Using the following formula, calculate this amount of rainfall in inches.

   0.394 (the number of inches in a centimeter)
   × 347 cm
   ____
   ? in.

INTERPRETING GRAPHICS

The graphs below show the monthly temperatures and rainfall in a region during 1 year.

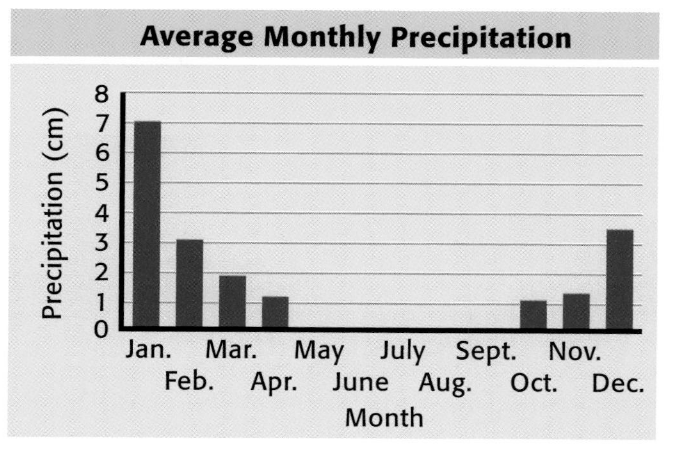

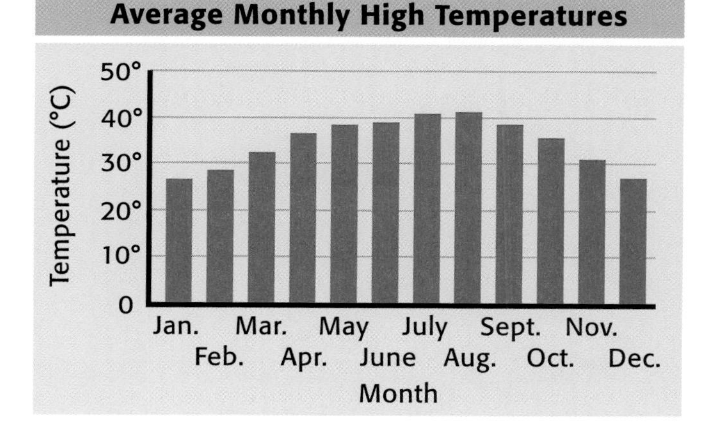

22. What kind of biome is probably found in the region represented by these graphs?

23. Would you expect to find bulrushes in the region represented by these graphs? Why or why not?

Take a minute to review your answers to the Pre-Reading Questions found at the bottom of page 46. Have your answers changed? If necessary, revise your answers based on what you have learned since you began this chapter.

Ocean Vents

▲ *"They're very slim, fuzzy, flattened-out worms. Really hairy," says scientist Bob Feldman about tubeworms.*

Picture the extreme depths of the ocean. There is no light at all, and it is very cold. But in the cracks between the plates on the bottom of the ocean floor, sea water trickles deep into the Earth. On the way back up from these cracks, the heated water collects metals, sulfuric gases, and enough heat to raise the temperature of the chilly ocean to 360°C. That is hot enough to melt lead! This heated sea water blasts up into the ocean through volcanic vents. And when this hot and toxic brew collides with icy ocean waters, the metals and sulfuric gases *precipitate,* that is, settle out of the heated ocean water as solids.

These solids form tubes, called black smokers, that extend up through the ocean floor. To humans, this dark, cold, and toxic environment would be deadly. But to a community of 300 species, including certain bacteria, clams, mussels, and tube worms, it is home. For these species, black smokers make life possible.

Life Without Photosynthesis

For a long time, scientists believed that energy from sunlight was the basis for the Earth's food chains and for life itself. But in the last 15 years, researchers have discovered ecosystems that challenge this belief. We now know of organisms around black smokers that can live without sunlight. One type of bacteria uses toxic gases from a black smoker in the same way that plants use sunlight. In a process called *chemosynthesis,* these bacteria convert sulfur into energy.

These bacteria are producers, and the mussels and clams are the consumers in this deep-sea food web. The bacteria use the mussels and clams as a sturdy place to live. The mussels and clams, in turn, feed off the bacteria. This kind of relationship between organisms is called *symbiosis.* The closer to the vent the clams and mussels are, the more likely the bacteria are to grow. Because of this, the mussels and clams frequently move to find good spots near the black smokers.

What Do You Think?

► Conditions near black smokers are similar to conditions on other planets. Do some research on these extreme environments, both on Earth and elsewhere. Then discuss with your classmates where and how you think life on Earth may have started.

CAREERS

ECOLOGIST

Most winters **Alfonso Alonso-Mejía** climbs up to the few remote sites in central Mexico where about 150 million monarch butterflies spend the winter. He is researching the monarchs because he wants to help preserve their habitat.

Monarch butterflies are famous for their long-distance migration. Those that eventually find their way to Mexico come from as far away as the northeastern United States and southern Canada. Some of them travel 3,200 km before reaching central Mexico.

Human Threats to Habitats

Unfortunately, the monarchs' habitat is increasingly threatened by logging and other human activities. Only nine of the monarchs' wintering sites remain. Five of the sites are set aside as sanctuaries for the butterflies, but even those are endangered by people who cut down fir trees for firewood or for commercial purposes.

Research to the Rescue

Alonso-Mejía's work is helping Mexican conservationists better understand and protect monarch butterflies. Especially important is his discovery that monarchs depend on bushlike vegetation that grows beneath the fir trees, called understory vegetation.

Alonso-Mejía's research showed that when the temperature dips below freezing, as it often does at the high-altitude sites where the monarchs winter, some monarchs depend on understory vegetation for survival. This is because low temperatures (−1°C to 4°C) limit the monarchs' movement—the butterflies are not even able to crawl. At extremely cold temperatures (−7°C to −1°C), monarchs resting on the forest floor are in danger of freezing to death. But where there is understory vegetation, the monarchs can slowly climb the vegetation until they are at least 10 cm above the ground. This tiny difference in elevation can provide a microclimate that is warm enough to ensure the monarchs' survival.

The importance of understory vegetation was not known before Alonso-Mejía did his research. Now, thanks to his work, Mexican conservationists will better protect the understory vegetation.

Get Involved!

► If you are interested in a nationwide tagging program to help scientists learn more about the monarchs' migration route, write to Monarch Watch, Department of Entomology, 7005 Howorth Hall, University of Kansas, Lawrence, Kansas 66045.

CHAPTER 4

Environmental Problems and Solutions

Sections

Pre-Reading Questions

1. Name three ways people damage the Earth.
2. Name three ways people are trying to prevent further damage to the Earth.

Sea of Red

On the shores of a tropical island, a chemical slick coats the clear water. Even though the environment may break down the chemical over time, preventing pollution is better for the environment. In this chapter, you will learn about the natural cycles that help break down substances and move them through the environment. You will also learn about problems facing the environment today and about some solutions, including actions you can take to help.

START-UP Activity

RECYCLING PAPER

In this activity, you will be making paper without cutting down trees. Instead you will be reusing paper that has already been made.

Procedure

1. Tear up **two sheets of old newspaper** into small pieces, and put them in a **blender.** Using a **beaker,** add **1 L of water.** Cover and blend until the mixture is soupy.
2. Cover the bottom of a **square pan** with **2-3 cm of water.** Place a **wire screen** in the pan. Pour 250 mL of the paper mixture onto the screen, and spread evenly.
3. Lift the screen out of the water with the paper on it. Drain excess water into the pan. Then place the screen inside a **section of newspaper.**
4. Close the newspaper, and turn it over so that the screen is on top of the paper mixture. Cover the newspaper with a **flat board.** Press on the board to squeeze out extra water.
5. Open the newspaper, and let your paper mixture dry. Use your recycled paper to write a note to a friend!

Analysis

6. In what ways is your paper like regular paper? How is it different?
7. What could you do to improve your papermaking methods?

SECTION 1

READING WARM-UP

Terms to Learn

pollution
renewable resource
nonrenewable resource
overpopulation
biodiversity
biodegradable

What You'll Do

- Describe the major types of pollution.
- Distinguish between renewable and nonrenewable resources.
- Explain how habitat destruction affects organisms.
- Explain the impact of human population growth.

First the Bad News

You've probably heard it before. The air is unhealthy to breathe. The water is harmful to drink. The soil is filled with poisons. The message is that Earth is sick and in great danger.

Pollution

Pollution is the presence of harmful substances in the environment. These harmful substances, known as *pollutants,* take many forms. They may be solid materials, chemicals, noise, or even heat. Often, pollutants damage or kill the plants and animals living in the affected habitat, as shown in **Figure 1.** Pollutants may also harm humans.

Figure 1 *The water poured into the river by this factory is polluted with chemicals and heat. The smoke contains harmful chemicals that pollute the air.*

Piles of Garbage Americans produce more household waste than any other nation. If stacked up, the beverage cans we use in one year could reach the moon 17 times! The average American throws away 12 kg of trash a week, which usually winds up in a landfill like the one in **Figure 2.** Businesses, mines, and industries also produce large amounts of wastes.

Billions of kilograms of this waste are classified as *hazardous waste,* which means it's harmful to humans and the environment. Many industries produce hazardous wastes, including paper mills, nuclear power plants, oil refineries, and plastic and metal processing plants. Hospitals and laboratories produce hazardous medical wastes. But industry shouldn't get all the blame. Hazardous wastes also come from homes. Old cars, paints, batteries, medical wastes, and detergents all pollute the environment.

Figure 2 *Every year, we throw away 150 million metric tons of garbage.*

Where Does It All Go? Most of our household waste goes into giant landfills. Hazardous wastes are buried in landfills specially designed to contain them. However, some companies illegally dispose of their hazardous wastes by dumping them into rivers and lakes. Some wastes are burned in incinerators designed to reduce the amount of pollutants that enter the atmosphere. But if wastes are burned improperly, they add to the pollution of the air.

Figure 3 *Rachel Carson's book* Silent Spring, *published in 1962, made people aware of the environmental dangers of pesticides, especially to birds.*

Chemicals Are Everywhere Chemicals are used to treat diseases. They are also used in plastics, thermometers, paints, hair sprays, and preserved foods. In fact, chemicals are everywhere. We can't get along without chemicals. Sometimes, though, we cannot get along *with* them. Chemical pesticides used to kill crop-destroying insects also pollute the soil and water. Rachel Carson, shown in **Figure 3,** wrote about the dangers of pesticides more than three decades ago.

A class of chemicals called CFCs was once used in aerosol sprays, refrigerators, and plastics. These uses of CFCs have been banned. CFCs rise high into the atmosphere and can cause the destruction of ozone. Ozone protects the Earth from harmful ultraviolet light.

Another class of chemicals, called PCBs, was once used as insulation as well as in paints, household appliances, and other products. Then scientists learned that PCBs are *toxic,* or poisonous. PCBs are now banned, but they have not gone away. They break down very slowly in the environment, and they still pollute even the most remote areas on Earth, as shown in **Figure 4.**

High-Powered Wastes Nuclear power plants produce electricity for millions of homes and businesses. They also produce *radioactive wastes,* special kinds of hazardous wastes that take hundreds or thousands of years to become harmless. These "hot" wastes can cause cancer, leukemia, and birth defects in humans. Radioactive wastes can have harmful effects on all living things.

Figure 4 *PCBs and other pollutants have even been found in remote parts of the Arctic.*

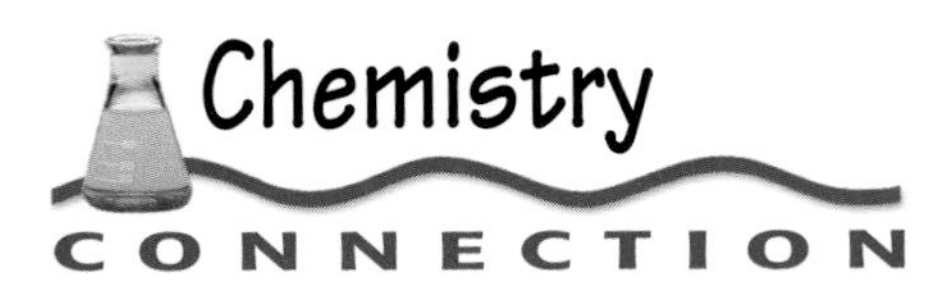

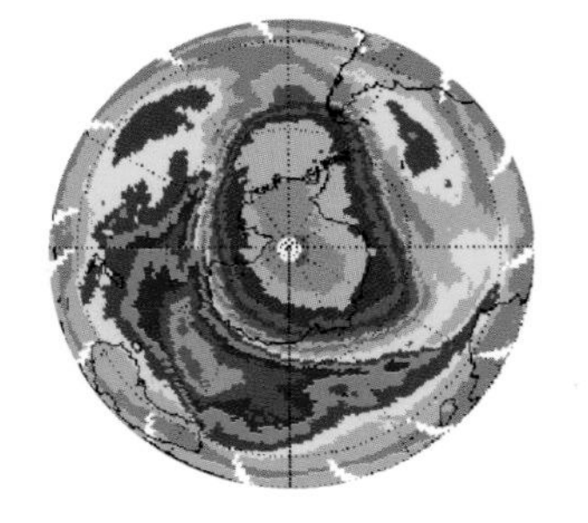

Ozone in the stratosphere absorbs most of the ultraviolet light that comes from the sun. Ozone is destroyed by CFCs. This image of the hole in the ozone layer (the gray area in the center) was taken in 1998.

Exposure to high levels of ultraviolet light can lead to blindness, rapid skin aging, skin cancer, and a weakened immune system.

Too Much Heat The Earth is surrounded by a mixture of gases, including carbon dioxide, that make up the atmosphere. The atmosphere acts as a protective blanket, keeping the Earth warm enough for life to exist. Since the late 1800s, however, the amount of carbon dioxide in the air has increased by 25 percent. Carbon dioxide and certain pollutants in the air act like a greenhouse. Most scientists think the increase in carbon dioxide and other pollutants has caused a significant increase in global temperatures. If the temperatures continue to rise, the polar icecaps could melt, raising the level of the world's oceans. Some scientists think the sea level could rise 10 cm to 1.2 m by the year 2100. A 1 m rise would flood coastal areas, pollute underground water supplies, and cause present shorelines to disappear.

It's Way Too Noisy! Some pollutants affect the senses. These include bad odors and loud noises. Too much noise is not just annoying; it affects the ability to hear and think. If construction workers and others who work in noisy environments do not protect their ears, they can slowly lose their hearing. The students shown in **Figure 5** are listening to music at a sensible volume so that their hearing will not be damaged.

Figure 5 *Listening to music at a sensible volume will help prevent hearing loss.*

internet**connect**

TOPIC: Air Pollution
GO TO: www.scilinks.org
***sci*LINKS NUMBER:** HSTL505

SECTION REVIEW

1. Describe two ways pollution can be harmful.
2. Explain how loud noise can be considered pollution.
3. **Applying Concepts** Explain how each of the following can help people but harm the environment: hospitals, refrigerators, and road construction.

Resource Depletion

Another problem for our environment is that we are using up, or depleting, natural resources. Some of the Earth's resources are renewable, but others are nonrenewable. A **renewable resource** is one that can be used again and again or has an unlimited supply. Fresh water and solar energy are renewable resources, as are some kinds of trees. A **nonrenewable resource** is one that can be used only once. Most minerals are nonrenewable. Fossil fuels, such as oil and coal, are also nonrenewable resources.

Some nonrenewable resources, such as petroleum, are probably not in danger of running out in your lifetime. But we use more and more nonrenewable resources every year, and they cannot last forever. Plus, the removal of some materials from the Earth carries a high price tag in the form of oil spills, loss of habitat, and damage from mining, as shown in **Figure 6.**

Water Depletion

An underground water supply has a depth of 200 m of water. Water seeps in at the rate of 4 cm/year. Water is pumped out at the rate of 1 m/year. How long will this water supply last?

To find the net water loss from an underground water supply, subtract the amount that seeps into the water supply from the amount removed from the water supply.

How long will the water supply last if water seeps in at the rate of 10 cm/year and is removed at the rate of 10 cm/year?

Figure 6 *This area has been mined for coal using a method called strip mining.*

Nonrenewable or Renewable? Some resources once thought to be renewable are becoming nonrenewable. Ecosystems, such as tropical rain forests, are being polluted and destroyed, resulting in huge losses of habitat. Around the world, rich soil is being eroded away and polluted. A few centimeters of soil takes thousands of years to form and can be washed away in less than a year. Underground water needed for drinking and irrigation is used faster than it is replaced. Several centimeters of water may seep into an underground source each year, but in the same amount of time, *meters* of water are being pumped out.

1. In what ways do you use nonrenewable resources?
2. Why would it not be a good idea to use up a nonrenewable resource?

(See page 168 to check your answers.)

Figure 7 *The zebra mussel is an alien invader that is clogging water treatment plants in the Great Lakes region.*

Alien Species

People are constantly on the move. Without knowing it, we take along passengers. Boats, airplanes, and cars carry plant seeds, animal eggs, and adult organisms from one part of the world to another. An organism that makes a home for itself in a new place is an *alien.* One reason alien species often thrive in foreign lands is that they are free from the predators in their native habitats.

Alien species often become pests and drive out native species. The zebra mussel, shown in **Figure 7,** hitched a ride on ships sailing from Europe to the United States in the 1980s. The purple loosestrife, shown in **Figure 8,** arrived long ago from Europe. Today it is crowding out native vegetation and threatening rare plant species in much of North America. Many organisms, such as the dandelion, are so common and have been here so long that it is easy to forget they don't belong.

Figure 8 *The purple loosestrife from Europe is choking out natural vegetation in North America.*

Human Population Growth

In 1800, there were 1 billion people on Earth. In 1990, there were 5.2 billion. By 2100, there may be 14 billion. Today, one out of ten people goes to bed hungry every night, and millions die each year from hunger-related causes. Some people believe that the human population is already too high for the Earth to support.

More people require more resources, and the human population is growing rapidly. **Overpopulation** occurs when the number of individuals becomes so large that they can't get all the food, water, and other resources they need on an ongoing basis.

Figure 9 shows that it took most of human history for the human population to reach 1 billion. Will the planet be able to support 14 billion people?

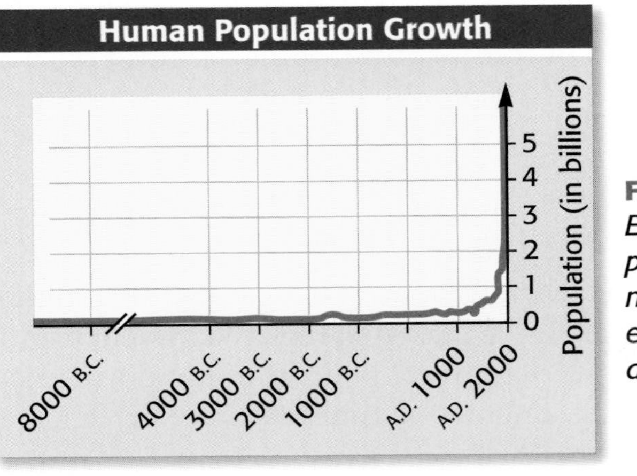

Figure 9 *The Earth's human population is now doubling every few decades.*

Habitat Destruction

The term **biodiversity** means "variety of life." It refers to the many different species found in a particular habitat all across the planet.

Every habitat has its own diverse combination of occupants. Every time a bulldozer digs or a chainsaw buzzes, every time hazardous wastes are dumped, a habitat is damaged, changed, or destroyed. And every time a habitat is destroyed, biodiversity is lost.

Figure 10 *Temperate forests are destroyed for many of the same reasons that tropical rain forests are destroyed.*

Forests Trees give us oxygen, furniture, fuel, fruits and nuts, rubber, alcohol, paper, turpentine, pencils, and telephone poles. Once trees covered twice as much land as they do today. *Deforestation,* such as that in **Figure 10,** is the clearing of forest lands. Tropical forests are cut for mines, dams, and roads. They are also cleared for paper, fuel, and building materials. But after tropical rain forests are cleared, little can grow on the land. Tropical soil doesn't have many nutrients, so it cannot be used for farming and is often abandoned.

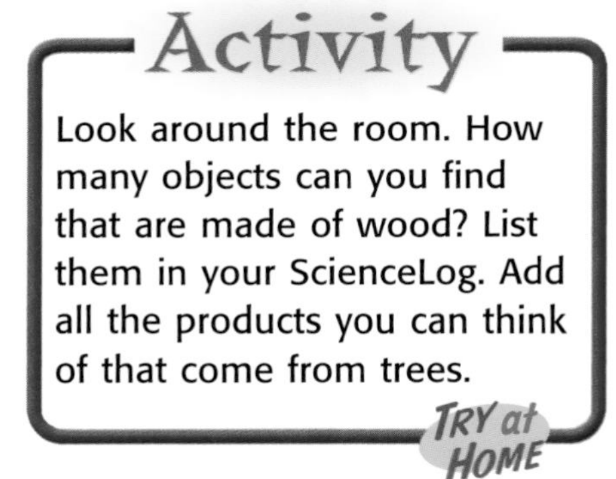

Activity

Look around the room. How many objects can you find that are made of wood? List them in your ScienceLog. Add all the products you can think of that come from trees.

TRY at HOME

Wetlands Wetlands were once considered unimportant. But as you know, that's not true. Wetlands help control flooding by soaking up the water from overflowing rivers. They filter pollutants from flowing water and provide breeding grounds for animals. They help prevent soil erosion and restore underground water supplies. Yet wetlands are often drained and filled to provide land for farms, homes, and shopping malls. They are dredged to keep passages open for ships and boats. Wetland habitats can also be destroyed by pollution.

Marine Habitats Oil is a major contributor to marine habitat loss. Oil from cities and industries is sometimes dumped into the ocean. Accidental spills and waste from oil tankers add more oil to the oceans. Spilled oil contaminates both open waters and coastal habitats, as shown in **Figure 11.** All the oceans are connected, so pollutants from one ocean can be carried around the world.

Figure 11 *Oil from the* Exxon Valdez *damaged more than 2,300 km^2 of the Alaskan coast.*

Balloons Aloft

Your town is about to celebrate its 200th birthday. A giant birthday party is planned. As part of the celebration, the town plans to release 1,000 helium balloons that say "Happy Birthday to Our Town." Why is this not a good idea? What can you do to convince town officials to change their plans?

Figure 12 *This sea bird has become entangled in a plastic six-pack holder.*

Plastics are often dumped into marine habitats. They are lightweight and float on the surface. They are not **biodegradable,** so they are not broken down by the environment. Animals, such as the bird in **Figure 12,** try to eat them and often get tangled in them and die. Dumping plastics into the ocean is against the law, but it is difficult to enforce.

Effects on Humans

Trees and sea creatures are not the only organisms affected by pollution, global warming, and habitat destruction. The damage we do to the Earth affects us too. Sometimes the effect is immediate. If you drink polluted water, you may immediately get sick or even die. But sometimes the damage is not apparent right away. Some chemicals cause cancers 20 or 30 years after a person is exposed to them. Your children or grandchildren may have to deal with depleted resources.

Anything that endangers other organisms will eventually endanger us too. Taking good care of the environment requires being concerned about what is happening right now. It also requires looking ahead to the future.

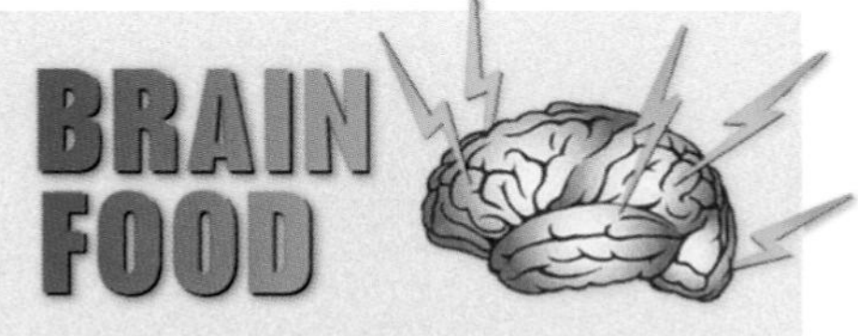

If humans became extinct, other organisms would go on living. But if all the insects became extinct, many plants could not reproduce. Animals would lose their food supply. The organisms we depend on, and eventually all of us, would disappear from the face of the Earth.

SECTION REVIEW

1. Why do alien species often thrive?
2. Explain how human population growth is related to pollution problems.
3. **Applying Concepts** How can the destruction of wetland habitats affect humans?

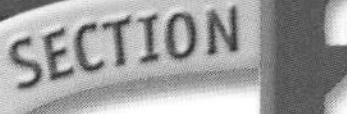

SECTION 2

READING WARM-UP

Terms to Learn

conservation
recycling
resource recovery

What You'll Do

- Explain the importance of conservation.
- Describe the three Rs and their importance.
- Explain how habitats can be protected.
- List ways you can help protect the Earth.

The Good News: Solutions

As you've seen, the news is bad. But it isn't *all* bad. In fact, there is plenty of good news. The good news is about what people can do—and are doing—to save the Earth. It is about what *you* can do to save the Earth. Just as people are responsible for damaging the Earth, people can also take responsiblity for helping to heal and preserve the Earth.

Conservation

One major way to help save the Earth is conservation. **Conservation** is the wise use of and the preservation of natural resources. If you ride your bike to your friend's house, you conserve fuel. At the same time, you prevent air pollution. If you use organic compost instead of chemical fertilizer on your garden, you conserve the resources needed to make the fertilizer. You also prevent soil and water pollution.

Practicing conservation means using fewer natural resources. It also means reducing waste. The three Rs, shown in **Figure 13,** describe three ways to conserve resources and reduce damage to the Earth: **R**educe, **R**euse, and **R**ecycle.

Figure 13 *These teenagers are observing the three Rs by using a cloth shopping bag, donating outgrown clothing to be reused, and recycling plastic.*

Reduce

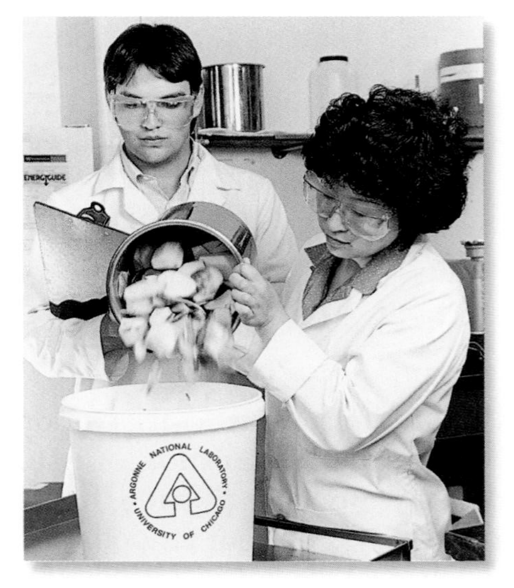

Figure 14 *These scientists are studying ways to use waste products to make biodegradable plastics.*

The most obvious way to conserve the Earth's resources is to use less. This will also help reduce pollution and wastes. Some companies have started using a variety of strategies to conserve resources. They often save money in the process.

Reducing Waste and Pollution One-third of the waste from cities and towns is packaging. To conserve resources and reduce waste, products can be wrapped in less paper and plastic. Fast foods can be wrapped in thin paper instead of large plastic containers that are not biodegradable. You can choose to take your purchases without a sack if you don't need one. Scientists, such as the ones in **Figure 14,** are working to make better biodegradable plastics.

Some companies are searching for less hazardous materials to use in making products. For example, some farmers refuse to use pesticides and chemical fertilizers. They practice organic farming. They use mulch, compost, manure, and natural pesticides. Agricultural specialists are also developing new farming techniques that are better for the environment.

Figure 15 *Rooftop solar panels provide most of the energy used in this neighborhood in Rotterdam, Holland.*

Reducing Use of Nonrenewable Resources Scientists are searching for alternative sources of energy. They want to avoid burning fuels and using nuclear energy. In some parts of the world solar energy heats water and powers homes, such as those shown in **Figure 15.** Engineers are working to make solar-powered cars practical. Other scientists are investigating the use of alternative power sources, such as wind, tides, and falling water.

It's Everyone's Responsibility Using fewer resources and reducing waste is not the job of industry and agriculture alone. Individuals use plenty of manufactured products and plenty of energy. They also produce large quantities of waste. Each United States citizen produces 40 times more waste than a citizen of a developing country. Why do you think this is so? What could you do to reduce the amount of trash that you produce? Everyone can take responsibility for helping to conserve the Earth's resources.

Reuse

Do you get hand-me-down clothes from an older sibling? Do you try to fix broken sports equipment instead of thowing it away? If so, you are helping preserve the Earth by *reusing* products.

Reusing Products Every time someone reuses a plastic bag, one less bag needs to be made, and one less bag pollutes the Earth. Every time someone uses a rechargeable battery, one less battery needs to be made, and one less battery will pollute the Earth. Reusing is an important way to conserve resources and prevent pollution.

Reusing Water About 85 percent of the water used in homes goes down the drain. Communities with water shortages are experimenting with reclaiming and reusing this waste water. Some use green plants or filter-feeding animals such as clams to clean the water. The water isn't pure enough to drink, but it is fine for watering lawns and golf courses, such as the one shown in **Figure 16.**

Figure 16 *This golf course is being watered with reclaimed water.*

Self-Check

1. How can you reduce the amount of electricity you use?
2. List five products that can be reused easily.

(See page 168 to check your answers.)

Recycle

Recycling is a form of reuse. **Recycling** requires breaking down trash and using it again. Sometimes recycled products are used to make the same kind of products. Sometimes they are made into different products. The park bench in **Figure 17** was made from plastic foam cups, hamburger boxes, and plastic bottles that once held detergent, yogurt, and margarine. All of the containers pictured in **Figure 18** can be easily recycled.

Figure 17 *This park bench is made of melted, remolded, and reused plastic.*

Figure 18 *These containers are examples of common household trash that can be recycled.*

Figure 19 *Each kind of recycled material is sorted into its own bin and then delivered to a recycling plant for processing.*

Recycling Trash Plastics, paper, aluminum cans, waste wood, glass, and cardboard are some examples of materials that can be recycled. Every week, half a million trees are needed to make Sunday newspapers. Recycling newspapers could save many trees. Recycling aluminum foil and cans saves 95 percent of the energy needed to change raw ore into aluminum. Glass makes up 8 percent of all our waste. It can be remelted to make new bottles and jars. Lead batteries can be recycled into new batteries.

Some cities, such as Austin, Texas, make recycling easy. Special containers for glass, plastic, aluminum, and paper are provided to each city customer. Each week trash to be recycled is collected in special trucks, such as the one shown in **Figure 19,** at the same time other waste is collected.

Recycling Resources Waste that can be burned can also be used to generate electricity in factories like the one shown in **Figure 20.** The process of transforming garbage to electricity is called **resource recovery.** The waste collected by all the cities and towns in the United States could produce about the same amount of electricity as 15 large nuclear power plants. Some companies are beginning to do this with their own waste. It saves them money, and it is responsible management.

Recycling is not difficult. Yet in the United States, only about 11 percent of the garbage is recycled. This compares with about 30 percent in Europe and 50 percent in Japan.

Figure 20 *A waste-to-energy plant can provide electricity to many homes and businesses.*

SECTION REVIEW

1. Define and explain *conservation.*
2. Describe the three main ways to conserve natural resources.
3. **Analyzing Relationships** How does conservation of resources also reduce pollution and other damage to the Earth?

Maintaining Biodiversity

Imagine a forest with just one kind of tree. If a disease hits that species, the entire forest might be wiped out. Now imagine a forest with 10 different kinds of trees. If a disease hits one kind of tree, nine different species will remain. Look at **Figure 21.** This field is growing a very important crop—cotton. But it is not very diverse. For the crop to thrive, the farmer must carefully manage the crop with weedkillers, pesticides, and fertilizers. Biodiversity helps to keep communities naturally stable.

How much biodiversity is in your part of the world? Investigate by turning to page 134 in your LabBook.

Figure 21 *What could happen if a cotton disease hits this cotton field? Biodiversity is low in fields of crops like this one.*

Species variety is also important because each species makes a unique contribution to an ecosystem. In addition, many species are important to humans. They provide many things, such as foods, medicines, natural pest control, beauty, and companionship, to name just a few.

Figure 22 *The California condor is returning from the verge of extinction thanks to careful captive breeding.*

Species Protection One way to maintain biodiversity is through the protection of individual species. In the United States, the Endangered Species Act is designed to do just that. Endangered organisms are included in a special list. The law forbids activities that would damage a plant or animal on the endangered species list. It also requires the development of programs to help endangered populations recover. Some endangered species are now increasing in number, such as the California condor in **Figure 22.**

Unfortunately, the process of getting a species on the endangered list takes a long time. Many new species need to be added to the list. Many species become extinct even before they are listed!

Figure 23 *Setting aside public lands for wildlife is one way to protect habitats.*

Habitat Protection Waiting until a species is almost extinct to begin protecting it is like waiting until your teeth are rotting to begin brushing them. Scientists want to prevent species from becoming endangered as well as from becoming extinct.

Plants, animals, and microorganisms do not live independently. Each is part of a huge interconnected web of organisms. To protect the entire web and to avoid disrupting the worldwide balance of nature, complete habitats, not just individual species, must be preserved. *All* species, not just those that are endangered, must be protected. All of the species that live in the nature preserve pictured in **Figure 23** are protected because the entire habitat is protected.

Strategies

Laws have been enacted to help conserve the Earth's environment. The purposes of such laws are listed below along with some of the ways citizens can help achieve these goals.

- **Reduce pesticide use.**
  Spray only pesticides that are targeted specifically for harmful insects. Use natural pesticides that interfere with the ways certain insects grow, develop, and live. Develop more biodegradable pesticides that will not injure birds, animals, or plants.
- **Reduce pollution.**
  Regulations prohibit the dumping of toxic substances and solid wastes into rivers, streams, lakes, and oceans and onto farmland and forests.
- **Protect habitats.**
  Conserve wetlands. Reduce deforestation. Practice logging techniques that consider the environment. Use resources at a rate that allows them to be replenished. Protect entire habitats.
- **Enforce the Endangered Species Act.**
  Speed up the process of getting endangered organisms listed.
- **Develop alternative energy sources.**
  Increase the use of solar power, wind power, and other renewable energy sources.

Trash Check

Keep track of all the trash you produce in one day. Classify it into groups. How much is food scraps? What might be considered a hazardous waste? What can be recycled? What can be reused? How can you reduce the amount of trash you produce?

TRY at HOME

What *You* Can Do

Reduce, reuse, recycle. Protect the Earth. These are jobs for everyone. Children as well as adults can help to save the Earth. The following list offers some suggestions for how *you* can help. How many of these things do you already do? What can you add to the list?

1. Buy things in packages that can be recycled.
2. Give away your old toys.
3. Use recycled paper.
4. Fill up both sides of a sheet of paper.
5. If you can't use permanent dishes, use paper instead of plastic-foam cups and plates.
6. Recycle glass, plastics, paper, aluminum, and batteries.
7. Don't buy anything made from an endangered animal.
8. Use rechargeable batteries.
9. Turn off lights, CD players, and computers when not in use.
10. Wear hand-me-downs.
11. Share books with friends, or use the library.
12. Walk, ride a bicycle, or use public transportation.
13. Carry a reusable cloth shopping bag to the store.
14. Use a lunch box or reuse your paper lunch bags.
15. Turn off the water while you brush your teeth.
16. Make a compost heap.
17. Buy products made from biodegradable plastic.
18. Use cloth napkins and kitchen towels.
19. Buy products with little or no packaging.
20. Repair leaking faucets.

SECTION REVIEW

1. Describe why biodiversity is important.
2. Why is it important to protect entire habitats?
3. **Applying Concepts** In the list above, identify which suggestions involve reducing, reusing, or recycling. Some suggestions will involve more than one of the three Rs.

Skill Builder Lab

Deciding About Environmental Issues

You make hundreds of decisions every day. Some of them are complicated, but many of them are very simple, such as what to wear or what to eat for lunch. Deciding what to do about an environmental issue can be more difficult. Many different factors must be considered. How will a certain solution affect people's lives? How much will it cost? Is it ethically right?

In this activity, you will analyze an issue in four steps to help you make a decision about it. Find out about environmental issues that are being discussed in your area. Examine newspapers, magazines, and other publications to find out what the issues are. Choose one local issue to evaluate. For example, you could evaluate whether the city should spend the money to provide recycling bins and special trucks for picking up recyclable trash.

MATERIALS

- newspapers, magazines, and other publications containing information about environmental issues

Procedure

1. In your ScienceLog, write a statement about an environmental issue.

2. *Gather Information* Read about your issue in several publications. Summarize important facts in your ScienceLog.

3. *Consider Values* Values are the things that you consider important. Examine the diagram below. Several values are given. Which values do you think apply most to the environmental issue you are considering? Are there other values that you believe will help you make a decision about the issue? Consider at least four values in making your decision.

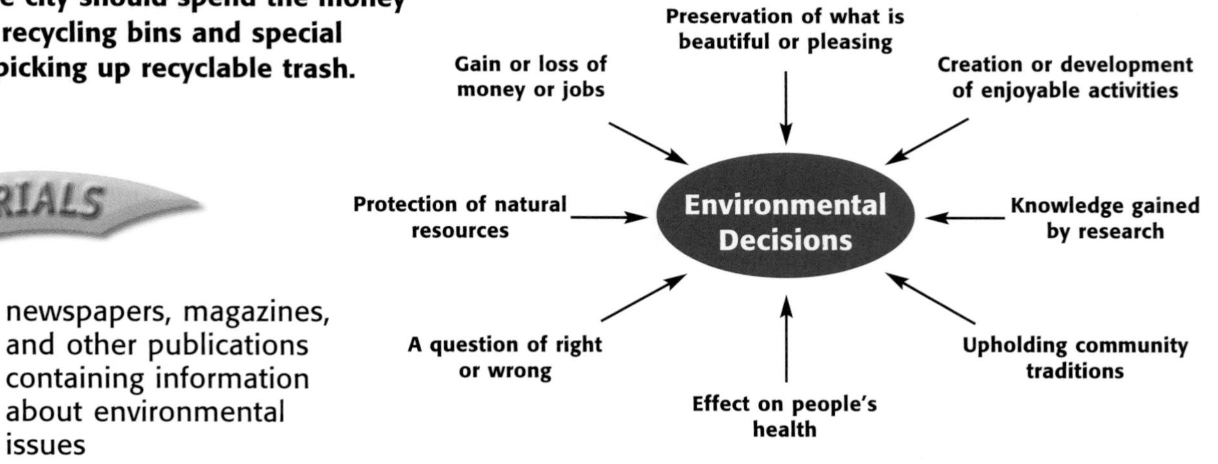

Consequences				
Types of consequences	**Value 1:**	**Value 2:**	**Value 3:**	**Value 4:**
Positive short-term consequences				
Negative short-term consequences				
Positive long-term consequences				
Negative long-term consequences				

4. *Explore Consequences* Consequences are the things that result from a certain course of action. In your ScienceLog, create a table similar to the one above. Use your table to organize your thoughts about consequences related to your environmental issue. List your values at the top. Fill in each space with the consequences of an action related to each of your values.

5. *Make a Decision* Thoroughly consider all of the consequences you have recorded in your table. Evaluate how important each consequence is. Make a decision about what course of action you would choose on the issue.

Analysis

6. In your evaluation, did you consider short-term consequences or long-term consequences to be more important? Why?

7. Which value or values had the greatest influence on your final decision? Explain your reasoning.

Going Further

Compare your table with your classmates' tables. Did you all make the same decision about a similar issue? If not, form teams and organize a formal classroom debate on a specific environmental issue.

Chapter Highlights

SECTION 1

Vocabulary

pollution *(p. 74)*
renewable resource *(p. 77)*
nonrenewable resource *(p. 77)*
overpopulation *(p. 78)*
biodiversity *(p. 79)*
biodegradable *(p. 80)*

Section Notes

- The Earth is being polluted by solid wastes, hazardous chemicals, radioactive materials, noise, and heat.
- Some of the Earth's resources renew themselves, and others do not. Some of the nonrenewable resources are being used up.
- Alien species often invade foreign lands, where they may thrive, become pests, and threaten native species.
- The human population is in danger of reaching numbers that the Earth cannot support.
- The Earth's habitats are being destroyed in a variety of ways, including deforestation, the filling of wetlands, and pollution.
- Deforestation may cause the extinction of species and often leaves the soil infertile.
- Air, water, and soil pollution can damage or kill animals, plants, and microorganisms.
- Humans depend on many different kinds of organisms. Pollution, global warming, habitat destruction—anything that affects other organisms will eventually affect humans too.

☑ Skills Check

Math Concepts

NET WATER LOSS Suppose that water seeps into an underground water supply at the rate of 10 cm/year. The underground water supply is 100 m deep, but it is being pumped out at about 2 m/year. How long will the water last?

First convert all measurements to centimeters.

(100 m = 10,000 cm; 2 m = 200 cm)

Then find the net loss of water per year.

200 cm – 10 cm = 190 cm (net loss per year)

Now divide the depth of the underground water supply by the net loss per year to find out how many years this water supply will last.

10,000 cm ÷ 190 cm = 52.6 years

Visual Understanding

THINGS YOU CAN DO Obviously, the strategies listed on page 86 to help preserve the Earth's habitats are strategies that scientists and other professionals are developing. To help you understand some of the things that you can do now, review the list on page 87.

SECTION 2

Vocabulary

conservation *(p. 81)*
recycling *(p. 83)*
resource recovery *(p. 84)*

Section Notes

- Conservation is the wise use of and preservation of the Earth's natural resources. By practicing conservation, people can reduce pollution and ensure that resources will be available to people in the future.
- Conservation involves the three Rs: Reduce, Reuse, and Recycle. Reducing means using fewer resources to begin with. Reusing means using materials and products over and over. Recycling involves breaking down used products and making them into new ones.
- Biodiversity is the variety of life on Earth. It is vital for maintaining stable, healthy, and functioning ecosystems.
- Habitats can be protected by using fewer pesticides, reducing pollution, avoiding habitat destruction, protecting species, and using alternative renewable sources of energy.
- Everyone can help to save the Earth by practicing the three Rs in their daily life.

Labs

Biodiversity—What a Disturbing Thought! *(p. 134)*

internet**connect**

GO TO: go.hrw.com

Visit the **HRW** Web site for a variety of learning tools related to this chapter. Just type in the keyword:

KEYWORD: HSTENV

GO TO: www.scilinks.org

Visit the **National Science Teachers Association** on-line Web site for Internet resources related to this chapter. Just type in the ***sci*LINKS** number for more information about the topic:

TOPIC: Air Pollution — ***sci*LINKS NUMBER:** HSTL505
TOPIC: Resource Depletion — ***sci*LINKS NUMBER:** HSTL510
TOPIC: Population Growth — ***sci*LINKS NUMBER:** HSTL515
TOPIC: Recycling — ***sci*LINKS NUMBER:** HSTL520
TOPIC: Maintaining Biodiversity — ***sci*LINKS NUMBER:** HSTL525

Chapter Review

USING VOCABULARY

To complete the following sentences, choose the correct term from each pair of terms listed below:

1. __?__ is the presence of harmful substances in the environment. *(Pollution* or *Biodiversity)*

2. __?__ is a type of pollution produced by nuclear power plants. *(CFC* or *Radioactive waste)*

3. A __?__ resource can be used only once. *(nuclear* or *nonrenewable)*

4. __?__ is the variety of forms among living things. *(Biodegradable* or *Biodiversity)*

5. __?__ is the breaking down of trash and using it to make a new product. *(Recycling* or *Reuse)*

UNDERSTANDING CONCEPTS

Multiple Choice

6. Habitat protection is important because
   a. organisms do not live independently.
   b. protecting habitats is a way to protect species.
   c. without it the balance of nature could be disrupted.
   d. All of the above

7. The Earth's resources can be conserved
   a. only by the actions of industry.
   b. by reducing the use of nonrenewable resources.
   c. if people do whatever they want to do.
   d. by throwing away all our trash.

8. Endangered species
   a. are those that are extinct.
   b. are found only in tropical rain forests.
   c. can sometimes be brought back from near extinction.
   d. are all protected by the Endangered Species Act.

9. Global warming is a danger
   a. only to people living in warm climates.
   b. to organisms all over the planet.
   c. only to life at the poles.
   d. to the amount of carbon dioxide in the air.

10. Overpopulation
    a. does not occur among human beings.
    b. helps keep pollution levels down.
    c. occurs when a species cannot get all the food, water, and other resources it needs.
    d. occurs only in large cities.

11. Biodiversity
    a. is of no concern to scientists.
    b. helps to keep ecosystems stable.
    c. causes diseases to destroy populations.
    d. is found only in temperate forests.

Short Answer

12. Describe how you can help to conserve resources. Include strategies from all of the three Rs.

13. Describe the connection between alien species and endangered species.

Concept Mapping

14. Use the following terms to create a concept map: pollution, pollutants, CFCs, cancer, PCBs, toxic, radioactive wastes, global warming.

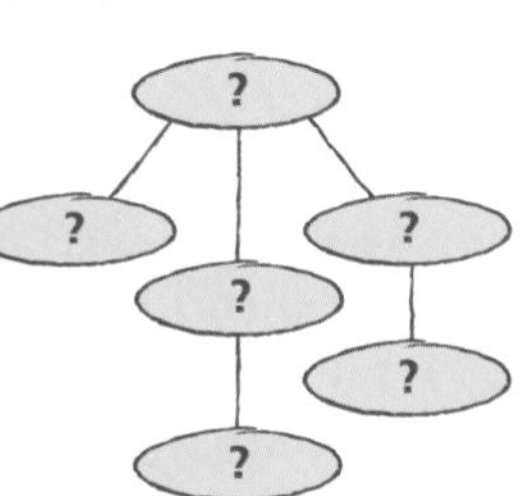

CRITICAL THINKING AND PROBLEM SOLVING

Write one or two sentences to answer the following question:

15. Suppose that the supply of fossil fuels were going to run out in 10 years. What would happen if we ran out without being prepared? What could be done to prepare for such an event?

MATH IN SCIENCE

16. If each person in a city of 150,000 throws away 12 kg of trash every week, how many metric tons of trash does the city produce per year? (There are 52 weeks in a year and 1,000 kg in a metric ton.)

INTERPRETING GRAPHICS

The illustration above shows how people in one home use natural resources.

17. Identify ways in which the people in this picture are wasting natural resources. Describe at least three examples, and tell what could be done to conserve resources.

18. Identify which resources in this picture are renewable.

19. Identify any sources of hazardous waste in this picture.

20. Explain how the girl wearing headphones is reducing pollution in the air. How could such a choice cause her harm?

Take a minute to review your answers to the Pre-Reading Questions found at the bottom of page 72. Have your answers changed? If necessary, revise your answers based on what you have learned since you began this chapter.

BIOLOGIST

Dagmar Werner works at the Carara Biological Preserve, in Costa Rica, protecting green iguanas. Suffering from the effects of hunting, pollution, and habitat destruction, the iguana was nearly extinct. Since the 1980s, Werner has improved the iguanas' chances of survival by breeding them and releasing thousands of young iguanas into the wild. She also trains other people to do the same.

At Werner's "iguana ranch" preserve, female iguanas lay their eggs in artificial nests. After they hatch, the young lizards are placed in a temperature- and humidity-controlled incubator and given a special diet. As a result, the iguanas grow faster and stronger and are better protected from predators than their wild counterparts. Ordinarily, less than 2 percent of all iguanas survive to adulthood, but Werner's iguanas have an 80 percent survival rate. Once the iguanas are released into the wild, Werner tracks and monitors them to determine whether they have successfully adapted to their less-controlled environment.

Chicken of the Trees

Because she knew the iguana was close to extinction, Werner took an immediate and drastic approach to saving the lizards. She combined her captive-breeding program at the preserve with an education program that shows farmers a new way to make money from the rain forest. Instead of cutting down rain forest to raise cattle, Werner encourages farmers to raise iguanas. The iguanas can be released into the wild or sold for food. Known as "chicken of the trees," this lizard has been a favored source of meat among rain-forest inhabitants for thousands of years.

With Werner's methods, farmers can protect iguanas and still earn a living. But convincing farmers hasn't been easy. According to Werner, "Many locals have never thought of wild animals as creatures that must be protected in order to survive." To help, Werner has established the Fundación Pro Iguana Verde (the Green Iguana Foundation), which sponsors festivals and education seminars. These activities promote the traditional appeal of the iguana, increase pride in the animal, and heighten awareness about the iguana's economic importance.

Find Other Solutions

▶ The green iguana is just one animal that is nearing extinction in the rain forest. Research another endangered species and find out what scientists and local communities are doing to protect the species. Does it seem to be working?

▶ *A green Iguana at Carara Biological Preserve, in Costa Rica*

Where Should the Wolves Roam?

The U.S. Fish and Wildlife Service has listed the gray wolf as an endangered species throughout most of the United States and has devised a plan to reintroduce the wolf to Yellowstone National Park, central Idaho, and northwestern Montana. The goal is to establish a population of at least 100 wolves at each location. If the project continues as planned, wolves may be removed from the endangered species list by 2002. But some ranchers and hunters are uneasy about the plan, and some environmentalists and wolf enthusiasts think that the plan doesn't go far enough to protect wolves.

Does the Plan Risk Livestock?

Ranchers are concerned that the wolves will kill livestock. These losses could result in a tremendous financial burden to ranchers. There is a compensation program currently established that will pay ranchers if wolves kill their livestock. But this program will end if the wolf is removed from the endangered species list. Ranchers point out that the threat to their livestock will not end when the wolf is removed from the list. In fact, the threat will increase, but ranchers will no longer receive any compensation.

On the other hand, some biologists offer evidence that wolves living near areas with adequate populations of deer, elk, moose, and other prey do not attack livestock. In fact, fewer than five wolf attacks on livestock were reported between 1995 and 1997.

Are Wolves a Threat to Wildlife?

Many scientists believe that the reintroduction plan would bring these regions into ecological balance for the first time in 60 years. They believe that the wolves will eliminate old and weak elk, moose, and deer and help keep these populations from growing too large.

Hunters fear that the wolves will kill many of the game animals in these areas. They cite studies that say large game animal populations cannot survive hunting by both humans and wolves. Hunting plays a significant role in the economy of the western states.

Are the People Safe?

Some people fear that wolves will attack people. However, there has never been a documented attack on humans by healthy wolves in North America. Supporters say that wolves are shy animals that prefer to keep their distance from people.

Most wolf enthusiasts admit that there are places where wolves belong and places where wolves do not belong. They believe that these reintroduction zones offer places for wolves to thrive without creating problems.

What Do You Think?

► Some people argue that stories about "the big, bad wolf" give the wolf its ferocious reputation. Do you think people's fears are based on myth, or do you think that the wolf is a danger to people and livestock living in the reintroduction zones? Do some research and provide examples to support your opinion.

◄ *A Gray Wolf in Montana*

CHAPTER 5

Energy Resources

Sections

LIVING INSIDE YOUR TRASH?

Would you believe that this house is made from empty soda cans and old tires? Well, it is! Not only does this house use recycled materials, but it also saves Earth's energy resources. This house gets all its energy from the sun and uses rainwater for household activities. In this chapter, you will learn about what Earth's energy resources are and how we can conserve them.

1. List four nonrenewable resources.
2. On which energy resources do humans currently depend the most?
3. What is the difference between a solar cell and a solar panel?

Architect Mike Reynolds in front of The Castle *in Taos, New Mexico*

START-UP Activity

WHAT IS THE SUN'S FAVORITE COLOR?

Are some colors better than others at absorbing the sun's energy? If so, how might this relate to collecting solar energy? Try the following activity to answer these questions.

Procedure

1. Obtain **at least five balloons** that are the same size and shape. One of the balloons should be white, and one should be black.
2. Place **one large ice cube or several small cubes** in each balloon. Each balloon should contain the same amount of ice.
3. Line the balloons up on a flat, uniformly colored surface that receives direct sunlight. Make sure that all the balloons receive the same amount of sunlight and that the openings in the balloons are not facing directly toward the sun.
4. Keep track of how much time it takes for the ice to melt completely in each of the balloons. You can tell how much ice has melted in each balloon by pinching the balloon's opening and then gently squeezing the balloon.

Analysis

5. In which balloon did the ice melt first? Why?
6. What color would you paint a device used to collect solar energy?

SECTION 1

READING WARM-UP

Natural Resources

Terms to Learn

natural resource
renewable resource
nonrenewable resource
recycling

What You'll Do

- Determine how humans use natural resources.
- Contrast renewable resources with nonrenewable resources.
- Explain how humans can conserve natural resources.

Think of the Earth as a giant life-support system for all of humanity. The Earth's atmosphere, waters, and solid crust provide almost everything we need to survive. The atmosphere provides the air we need to breathe, maintains air temperatures, and produces rain. The oceans and other waters of the Earth provide food and needed fluids. The solid part of the Earth provides nutrients and minerals.

Interactions between the Earth's systems can cause changes in the Earth's environments. Organisms must adapt to these changes if they are to survive. Humans have found ways to survive by using natural resources to change their immediate surroundings. A **natural resource** is any natural substance, organism, or energy form that living things use. Few of the Earth's natural resources are used in their unaltered state. Most resources are made into products that make people's lives more comfortable and convenient, as shown in **Figure 1.**

Figure 1 *Lumber, gasoline, and electricity are all products that come from natural resources.*

Renewable Resources

Some natural resources are renewable. A **renewable resource** is a natural resource that can be used and replaced over a relatively short time. **Figure 2** shows two examples of renewable resources. Although many resources are renewable, humans often use them more quickly than they can be replaced. Trees, for example, are renewable, but humans are currently cutting trees down more quickly than other trees can grow to replace them.

Figure 2 *Fresh water and trees are just a few of the renewable resources available on Earth.*

Nonrenewable Resources

Not all of Earth's natural resources are renewable. A **nonrenewable resource** is a natural resource that cannot be replaced or that can be replaced only over thousands or millions of years. Examples of nonrenewable resources are shown in **Figure 3.** The amounts of nonrenewable resources on Earth are fixed with respect to their availability for human use. Once nonrenewable resources are used up, they are no longer available. Oil and natural gas, for example, exist in limited quantities. When these resources become scarce, humans will have to find other resources to replace them.

Figure 3 *Nonrenewable resources, such as coal and natural gas, can be replaced only over thousands or millions of years once they are used up.*

Renewable or Nonrenewable?

Find five products in your home that were made from natural resources. List the resource or resources from which each product was made. Label each resource as renewable or nonrenewable.

Are the products made from mostly renewable or nonrenewable resources? Are those renewable resources plentiful on Earth? Do humans use those renewable resources more quickly than the resources can be replaced? What can you do to help conserve nonrenewable resources and renewable resources that are becoming more scarce?

Figure 4 *You can recycle many household items to help conserve natural resources.*

Conserving Natural Resources

Whether the natural resources we use are renewable or nonrenewable, we should be careful how we use them. To conserve natural resources, we should try to use them only when necessary. For example, leaving the faucet running while brushing your teeth wastes clean water. Turning the faucet on only to rinse your brush saves a lot of water that you or others need for other uses.

Another way to conserve natural resources is to recycle, as shown in **Figure 4.** **Recycling** is the process by which used or discarded materials are treated for reuse. Recycling allows manufacturers to reuse natural resources when making new products. This in turn reduces the amount of natural resources that must be obtained from the Earth. For example, recycling aluminum cans reduces the amount of aluminum that must be mined from the Earth's crust to make new cans.

internet**connect**

sciLINKS NSTA

TOPIC: Natural Resources
GO TO: www.scilinks.org
***sci*LINKS NUMBER:** HSTE105

SECTION REVIEW

1. How do humans use most natural resources?
2. What is the difference between renewable and nonrenewable resources?
3. Name two ways to conserve natural resources.
4. **Applying Concepts** List three renewable resources not mentioned in this section.

Fossil Fuels

Terms to Learn

energy resource
fossil fuel
petroleum
natural gas
coal
strip mining
acid precipitation
smog

What You'll Do

- Classify the different forms of fossil fuels.
- Explain how fossil fuels are obtained.
- Identify problems with fossil fuels.
- List ways to deal with fossil-fuel problems.

Energy resources are natural resources that humans use to produce energy. There are many types of renewable and nonrenewable energy resources, and all of the energy released from these resources ultimately comes from the sun. The energy resources on which humans currently depend the most are fossil fuels. **Fossil fuels** are nonrenewable energy resources that form in the Earth's crust over millions of years from the buried remains of once-living organisms. Energy is released from fossil fuels when they are burned. There are many types of fossil fuels, which exist as liquids, gases, and solids, and humans use a variety of methods to obtain and process them. These methods depend on the type of fossil fuel, where the fossil fuel is located, and how the fossil fuel formed. Unfortunately, the methods of obtaining and using fossil fuels can have negative effects on the environment. Read on to learn about fossil fuels and the role they play in our lives.

Liquid Fossil Fuels—Petroleum

Petroleum, or crude oil, is an oily mixture of flammable organic compounds from which liquid fossil fuels and other products, such as asphalt, are separated. Petroleum is separated into several types of fossil fuels and other products in refineries, such as the one shown in **Figure 5.** Among the types of fossil fuels separated from petroleum are gasoline, jet fuel, kerosene, diesel fuel, and fuel oil.

Figure 5 *Fossil fuels and other products are separated from petroleum in a process called* fractionation. *In this process, petroleum is gradually heated in a tower so that different components boil and vaporize at different temperatures.*

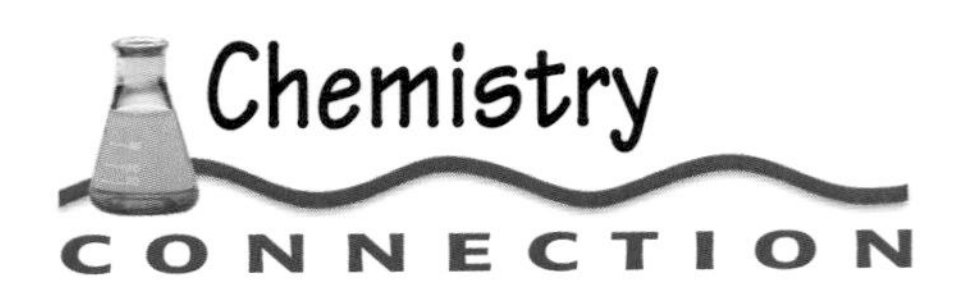

Petroleum and natural gas are both made of compounds called hydrocarbons. A ***hydrocarbon*** is an organic compound containing only carbon and hydrogen.

Gaseous Fossil Fuels—Natural Gas

Gaseous fossil fuels are classified as **natural gas.** Most natural gas is used for heating and for generating electricity. The stove in your kitchen may be powered by natural gas. Many motor vehicles, such as the van in **Figure 6,** are fueled by liquefied natural gas. Vehicles like these produce less air pollution than vehicles powered by gasoline.

Methane is the main component of natural gas. But other natural-gas components, such as butane and propane, can be separated and used by humans. Butane is often used as fuel for camp stoves. Propane is used as a heating fuel and as a cooking fuel, especially for outdoor grills.

Figure 6 *Vehicles powered by liquefied natural gas are becoming more common.*

Figure 7 *This coal is being gathered so that it may be burned in the power plant shown in the background.*

Solid Fossil Fuels—Coal

The solid fossil fuel that humans use most is coal. **Coal** is a solid fossil fuel formed underground from buried, decomposed plant material. Coal, the only fossil fuel that is a rock, was once the leading source of energy in the United States. People burned coal for heating and transportation. Many trains in the 1800s and early 1900s were powered by coal-burning steam locomotives.

People began to use coal less because burning coal often produces large amounts of air pollution and because better energy resources were discovered. Coal is no longer used much as a fuel for heating or transportation in the United States. However, many power plants, like the one shown in **Figure 7,** burn coal to produce electricity.

How Do Fossil Fuels Form?

All fossil fuels form from the buried remains of ancient organisms. But different types of fossil fuels form in different ways and from different types of organisms. Petroleum and natural gas form mainly from the remains of microscopic sea life. When these organisms die, their remains settle on the ocean floor, where they decay and become part of the ocean sediment. Over time, the sediment slowly becomes rock, trapping the decayed remains. Through physical and chemical changes over millions of years, the remains become petroleum and gas. Gradually, more rocks form above the rocks that contain the fossil fuels. Under the pressure of overlying rocks and sediments, the fossil fuels are squeezed out of their source rocks and into permeable rocks. As shown in **Figure 8,** these permeable rocks become reservoirs for petroleum and natural gas. The formation of petroleum and natural gas is an ongoing process. Part of the remains of today's sea life will probably become petroleum and natural gas millions of years from now.

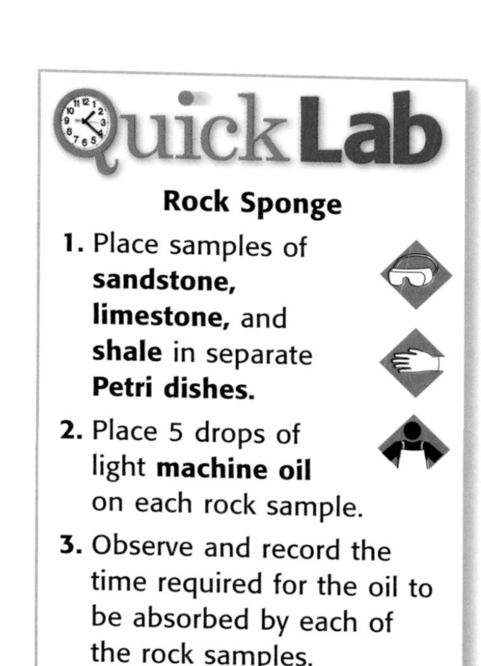

QuickLab

Rock Sponge

1. Place samples of **sandstone, limestone,** and **shale** in separate **Petri dishes.**
2. Place 5 drops of light **machine oil** on each rock sample.
3. Observe and record the time required for the oil to be absorbed by each of the rock samples.
4. Which rock sample absorbed the oil fastest? Why?
5. Based on your findings, describe a property that allows for easy removal of fossil fuels from reservoir rock.

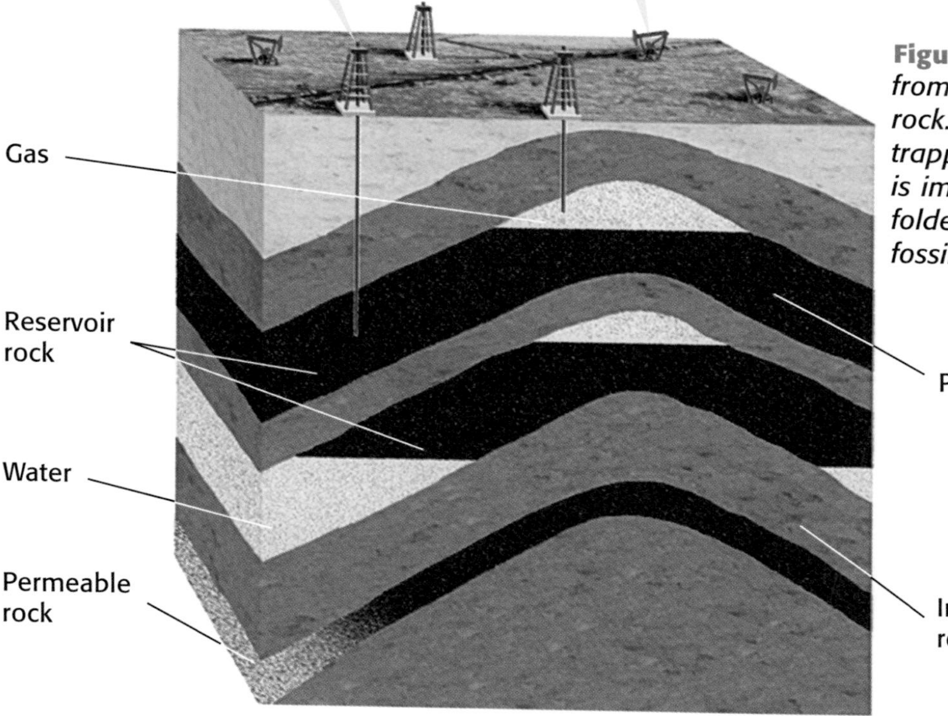

Figure 8 *Petroleum and gas rise from source rock into reservoir rock. Sometimes the fuels are trapped by overlying rock that is impermeable. Rocks that are folded upward are excellent fossil-fuel traps.*

Coal Formation Coal forms differently from petroleum and natural gas. Coal forms underground over millions of years from decayed swamp plants. When swamp plants die, they sink to the bottom of the swamps. This begins the process of coal formation, which is illustrated below. Notice that the percentage of carbon increases with each stage. The higher the carbon content, the cleaner the material burns. However, all grades of coal will pollute the air when burned.

The Process of Coal Formation

Stage 1: Peat
Bacteria and fungi transform sunken swamp plants into peat. Peat is about **60 percent carbon.**

Stage 2: Lignite
Sediment buries the peat, increasing the pressure and temperature. This gradually turns the peat into lignite, which is about **70 percent carbon.**

Stage 3: Bituminous coal
The temperature and pressure continue to increase. Eventually lignite turns into bituminous coal. Bituminous coal is about **80 percent carbon.**

Stage 4: Anthracite
With more heat and pressure, bituminous coal eventually turns into anthracite, which is about **90 percent carbon.**

SECTION REVIEW

1. Name a solid, liquid, and gaseous fossil fuel.
2. What component of coal-forming organic material increases with each step in coal formation?
3. **Comparing Concepts** What is the difference between the organic material from which coal forms and the organic material from which petroleum and natural gas mainly form?

Where Are Fossil Fuels Found?

Fossil fuels are found in many parts of the world, both on land and beneath the ocean. As shown in **Figure 9**, the United States has large reserves of petroleum, natural gas, and coal. In spite of all our petroleum reserves, we import about one-half of our petroleum and petroleum products from the Middle East, South America, and Africa.

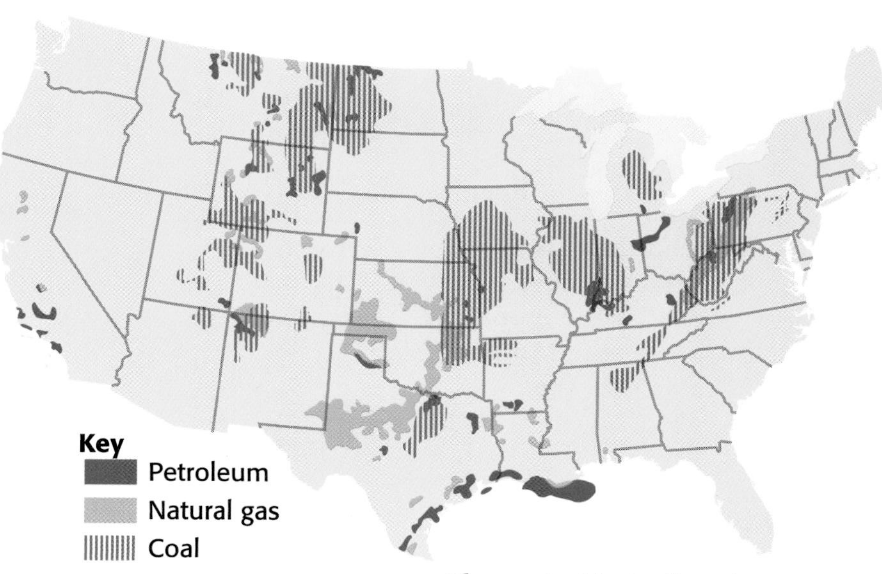

Figure 9 *Most oil and gas produced in the continental United States comes from California, Louisiana, and Texas.*

How Do Humans Obtain Fossil Fuels?

Humans use different methods to remove fossil fuels from the Earth's crust. These methods depend on the type of fuel being obtained and its location. Petroleum and natural gas are removed from the Earth by drilling wells into rock that contains these resources. Oil wells exist both on land and in the ocean. For offshore drilling, engineers mount drills on platforms that are secured to the ocean floor or float at the ocean's surface. **Figure 10** shows an offshore oil rig.

Coal is obtained either by mining deep beneath the Earth's surface or by strip mining. **Strip mining** is a process in which rock and soil are stripped from the Earth's surface to expose the underlying materials to be mined. Strip mining is used to mine shallow coal deposits. **Figure 11** shows a coal strip mine.

Figure 10 *Large oil rigs, some more than 300 m tall, operate offshore in many places, such as the Gulf of Mexico and the North Sea.*

Figure 11 *Strip miners use explosives to blast away rock and soil and to expose the material to be mined.*

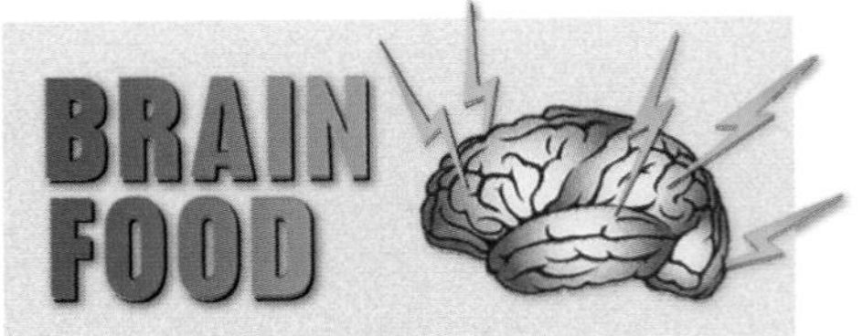

Although some countries have reduced their use of coal, the known coal reserves will last no more than 250 years at the present rates of coal consumption.

Problems with Fossil Fuels

Although fossil fuels provide energy for our technological world, the methods of obtaining and using them can have negative consequences. For example, when coal is burned, sulfur dioxide is released. Sulfur dioxide combines with moisture in the air to produce sulfuric acid, which is one of the acids in acid precipitation. **Acid precipitation** is rain or snow that has a high acid content due to air pollutants. Acid precipitation negatively affects wildlife, plants, buildings, and statues, as shown in **Figure 12.**

Figure 12 *Acid precipitation can dissolve parts of statues.*

Coal Mining The mining of coal can also create environmental problems. Strip mining removes soil, which plants need for growth and some animals need for shelter. If land is not properly repaired afterward, strip mining can destroy wildlife habitats. Coal mines that are deep underground, such as the one shown in **Figure 13,** can be hazardous to the men and women working in them. Coal mining can also lower local water tables, pollute water supplies, and cause the overlying earth to collapse.

Figure 13 *Coal dust can damage the human respiratory system. And because coal dust is flammable, it increases the danger of fire and explosion in coal mines.*

Petroleum Problems Obtaining petroleum can also cause environmental problems. In 1989, the supertanker *Exxon Valdez* spilled about 260,000 barrels of crude oil into the water when it ran aground off the coast of Alaska. The oil killed millions of animals and damaged the local fishing industry.

Smog Burning petroleum products causes a big environmental problem called smog. **Smog** is a photochemical fog produced by the reaction of sunlight and air pollutants. Smog is particularly serious in places such as Denver and Los Angeles. In these cities, the sun shines most of the time, there are millions of automobiles, and surrounding mountains prevent the wind from blowing pollutants away. Smog levels in some cities, including Denver and Los Angeles, have begun to decrease in recent years.

Dealing with Fossil-Fuel Problems

So what can be done to solve fossil-fuel problems? Obviously we can't stop using fossil fuels any time soon—we are too dependent on them. But there are things we can do to minimize the negative effects of fossil fuels. By traveling in automobiles only when absolutely necessary, people can cut down on car exhaust in the air. Carpooling, riding a bike, walking, and using mass-transit systems also help by reducing the number of cars on the road. These measures help reduce the negative effects of using fossil fuels, but they do not eliminate the problems. Only by using certain alternative energy resources, which you will learn about in the next section, can we eliminate them.

Figure 14 *Using mass transit, walking, or riding your bike can help reduce air pollution due to burning fossil fuels.*

SECTION REVIEW

1. Name a state with petroleum, natural-gas, and coal reserves.
2. How do we obtain petroleum and natural gas? How do we obtain coal?
3. Name three problems with fossil fuels. Name three ways to minimize the negative effects of fossil fuels.
4. **Making Inferences** Why does the United States import petroleum from other regions even though the United States has its own petroleum reserves?

internet**connect**

SCILINKS NSTA

TOPIC: Fossil Fuels
GO TO: www.scilinks.org
***sci*LINKS NUMBER:** HSTE120

SECTION 3

READING WARM-UP

Alternative Resources

Terms to Learn

nuclear energy
solar energy
wind energy
hydroelectric energy
biomass
gasohol
geothermal energy

What You'll Do

- Describe alternatives to the use of fossil fuels.
- List advantages and disadvantages of using alternative energy resources.

The energy needs of industry, transportation, and housing are increasingly met by electricity. However, most electricity is currently produced from fossil fuels, which are nonrenewable and cause pollution when burned. For people to continue their present lifestyles, new sources of energy must become available.

Splitting the Atom

Nuclear energy is an alternative source of energy that comes from the nuclei of atoms. Most often it is produced by a process called *fission*. Fission is a process in which the nuclei of radioactive atoms are split and energy is released, as shown in **Figure 15.** Nuclear power plants use radioactive atoms as fuel. When fission takes place, a large amount of energy is released. The energy is used to produce steam to run electric generators in the power plant.

Figure 15 *The process of fission generates a tremendous amount of energy.*

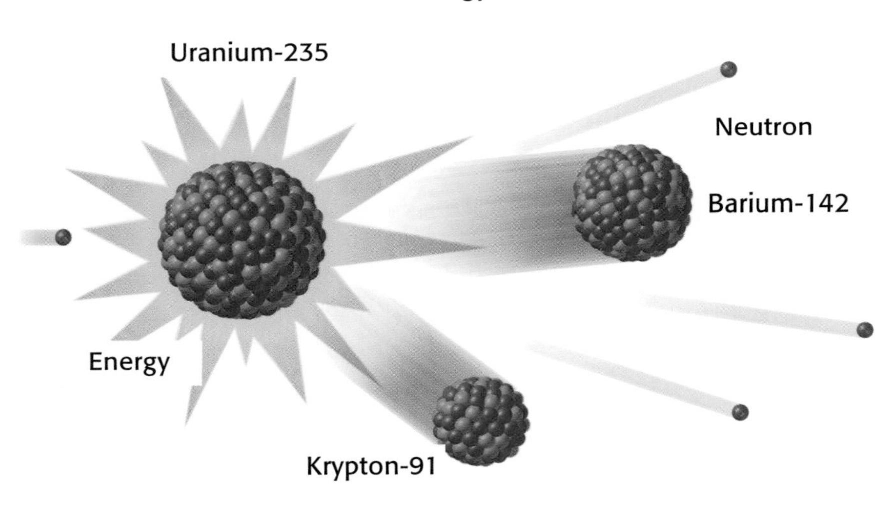

Pros and Cons Nuclear power plants provide alternative sources of energy without the problems that come with fossil fuels. So why don't we use nuclear energy instead of fossil fuels? Nuclear power plants produce dangerous wastes. The wastes are unsafe because they are radioactive. Radioactive wastes must be removed from the plant and stored until they lose their radioactivity. But nuclear wastes can remain dangerously radioactive for thousands of years. A safe place must be found to store these wastes so that radiation cannot escape into the environment.

Figure 16 *Areas or objects marked with this symbol should be approached only after taking proper precautions.*

Because nuclear power plants generate a lot of energy, large amounts of water are used in cooling towers, like the ones shown in **Figure 17**, to cool the plants. If a plant's cooling system were to stop working, the plant would overheat, and its reactor could possibly melt. Then a large amount of radiation could escape into the environment, as it did at Chernobyl, Ukraine, in 1986.

Figure 17 *Cooling towers are one of many safety mechanisms used in nuclear power plants. Their purpose is to prevent the plant from overheating.*

Combining Atoms

Another type of nuclear energy is produced by *fusion*. Fusion is the joining of nuclei of small atoms to form larger atoms. This is the same process that is thought to produce energy in the sun.

The main advantage of fusion is that it produces few dangerous wastes. The main disadvantage of fusion is that very high temperatures are required for the reaction to take place. No known material can withstand temperatures that high, so the reaction must occur within a special environment, such as a magnetic field. So far, fusion reactions have been limited to laboratory experiments.

Sitting in the Sun

When sunlight falls on your skin, the warmth you feel is part of solar energy. **Solar energy** is energy from the sun. Every day, the Earth receives more than enough solar energy to meet all of our energy needs. And because the Earth continuously receives solar energy, the energy is a renewable resource.

There are two common ways that we use solar energy. Sunlight can be changed into electricity by the use of solar cells. You may have used a calculator, like the one shown in **Figure 18**, that was powered by solar cells.

Figure 18 *This solar calculator receives all the energy it needs through the four solar cells located above its screen.*

Did you know that the energy from petroleum, coal, and natural gas is really a form of stored solar energy? All organisms ultimately get their energy from sunlight and store it in their cells. When ancient organisms died and became trapped in sediment, some of their energy was stored in the fossil fuel that formed in the sediment. So the gasoline that powers today's cars contains energy from sunlight that fell on the Earth millions of years ago!

Solar Cells A single solar cell produces only a tiny amount of electricity. For small electronic devices, such as calculators, this is not a problem because enough energy can be obtained with only a few cells. But in order to provide enough electricity for larger objects, such as a house, thousands of cells are needed. Many homes and businesses use solar panels mounted on their roof to provide much of their needed electricity. Solar panels are large panels made up of many solar cells wired together. **Figure 19** shows a building with solar panels.

Figure 19 *Although they are expensive to install, solar panels are good investments in the long run.*

Counting the Cost Solar cells are reliable and quiet, have no moving parts, and can last for years with little maintenance. They produce no pollution during use, and pollution created by their manufacturing process is very low.

So why doesn't everyone use solar cells? The answer is cost. While solar energy itself is free, solar cells are relatively expensive to make. The cost of a solar-power system could account for one-third of the cost of an entire house. But in remote areas where it is difficult and costly to run electric wires, solar-power systems can be a realistic option. In the United States today, tens of thousands of homes use solar panels to produce electricity. Can you think of other places that you have seen solar panels? Take a look at **Figure 20.**

Figure 20 *Perhaps you have seen solar panels used in this manner in your town.*

Solar Heating Another use of solar energy is direct heating through solar collectors. Solar collectors are dark-colored boxes with glass or plastic tops. A common use of solar collectors is heating water, as shown in **Figure 21.** Over 1 million solar water heaters have been installed in the United States. They are especially common in Florida, California, and some southwestern states.

As with solar cells, the problem with solar collectors is cost. But solar collectors quickly pay for themselves—heating water is one of the major uses of electricity in American homes. Also, solar collectors can be used to generate electricity.

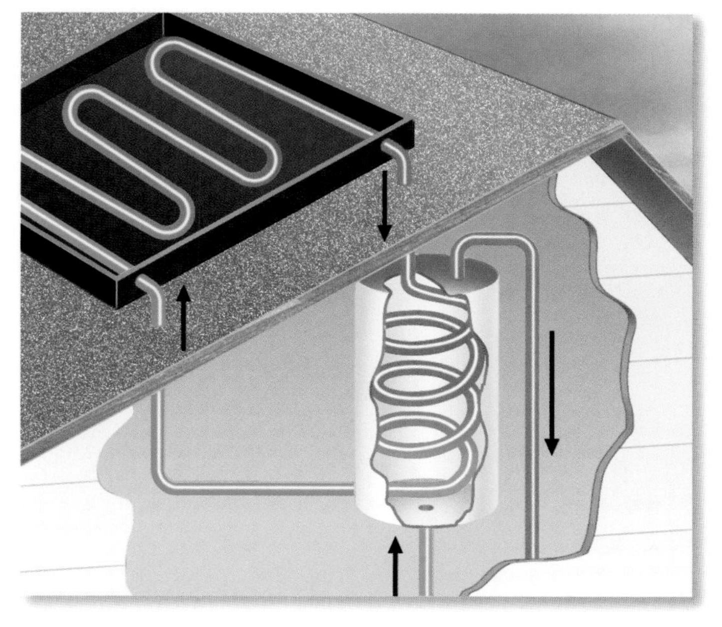

Figure 21 *After the liquid in the collector is heated by the sun, it is pumped through tubes that run through a water heater, causing the temperature of the water to rise.*

Large-Scale Solar Power Experimental solar-power facilities, such as the one shown in **Figure 22,** have shown that it is possible to generate electricity for an entire city. Facilities like this one are designed to use mirrors to focus sunlight onto coated steel pipes filled with synthetic oil. The oil is heated by the sunlight and is then used to heat water. The heated liquid water turns to steam, which is used to drive electric generators.

An alternative design for solar-power facilities is one that uses mirrors to reflect sunlight onto a receiver on a central tower. The receiver captures the sunlight's energy and stores it in tanks of molten salt. The stored energy is then used to create steam, which drives a turbine in an electric generator. *Solar Two,* a solar-power facility designed in this manner, is capable of generating enough energy to power 10,000 homes in southern California.

Turn to page 136 to calculate the power of the sun.

Figure 22 *This solar-power facility in the Mojave Desert has 1,926 sun-tracking mirrors called* heliostats.

Capture the Wind

Wind is created indirectly by solar energy through the uneven heating of air. There is a tremendous amount of energy in wind, called **wind energy.** You can see the effects of this energy unleashed in a hurricane or tornado. Wind energy can also be used productively by humans. Wind energy can turn a windmill that pumps water or produces electricity.

Figure 23 *Wind turbines take up only a small part of the ground's surface. This allows the land on wind farms to be used for more than one purpose.*

Wind Turbines Today, fields of modern wind turbines—technological updates of the old windmills—generate significant amounts of electricity. Clusters of these turbines are often called wind farms. Wind farms are located in areas where winds are strong and steady. Most of the wind farms in the United States are in California. The amount of energy produced by California wind farms could power all of the homes in San Francisco.

Steady Breezes There are many benefits of using wind energy. Wind energy is renewable. Wind farms can be built in only 3–6 months. Wind turbines produce no carbon dioxide or other air pollutants during operation. The land used for wind farms can also be used for other purposes, such as cattle grazing, as shown in **Figure 23.** However, the wind blows strongly and steadily enough to produce electricity on a large scale only in certain places. Currently, wind energy accounts for only a small percentage of the energy used in the United States.

internet**connect**

TOPIC: Renewable Resources
GO TO: www.scilinks.org
***sci*LINKS NUMBER:** HSTE110

SECTION REVIEW

1. Briefly describe two ways of using solar energy.
2. In addition to multiple turbines, what is needed to produce electricity from wind energy on a large scale?
3. **Analyzing Methods** Nuclear power plants are rarely found in the middle of deserts or other extremely dry areas. If you were going to build a nuclear plant, why would you not build it in the middle of a desert?

Hydroelectric Energy

The energy of falling water has been used by humans for thousands of years. Water wheels, such as the one shown in **Figure 24,** have been around since ancient times. In the early years of the Industrial Revolution, water wheels provided energy for many factories. More recently, the energy of falling water has been used to generate electricity. Electricity produced by falling water is called **hydroelectric energy.**

Figure 24 *Falling water turns water wheels, which turn giant millstones used to grind grain into flour.*

Harnessing the Water Cycle Hydroelectric energy is inexpensive and produces little pollution, and it is renewable because water constantly cycles from the ocean to the air, to the land, and back to the ocean. But like wind energy, hydroelectric energy is not available everywhere. Hydroelectric energy can be produced only where large volumes of falling water can be harnessed. Huge dams, like the one in **Figure 25,** must be built on major rivers to capture enough water to generate significant amounts of electricity.

Figure 25 *Falling water turns huge turbines inside hydroelectric dams, generating electricity for millions of people.*

At What Price? Increased use of hydroelectric energy could reduce the demand for fossil fuels, but there are trade-offs. Construction of the large dams necessary for hydroelectric power plants often destroys other resources, such as forests and wildlife habitats. For example, hydroelectric dams on the Lower Snake and Columbia Rivers in Washington disrupt the migratory paths of local populations of salmon and steelhead. Large numbers of these fish die each year because their life cycle is disrupted. Dams can also decrease water quality and create erosion problems.

Self-Check

How are ancient water wheels like modern hydroelectric dams? *(See page 168 to check your answer.)*

Powerful Plants

Plants are similar to solar collectors, absorbing energy from the sun and storing it for later use. Leaves, wood, and other parts of plants contain the stored energy. Even the dung of plant-grazing animals is high in stored energy. These sources of energy are called biomass. **Biomass** is organic matter that contains stored energy.

Burning Biomass Biomass energy can be released in several ways. The most common is the burning of biomass. Approximately 70 percent of people living in developing countries heat their homes and cook their food by burning wood or charcoal. In the United States this number is about 5 percent. Scientists estimate that the burning of wood and animal dung accounts for approximately 14 percent of the world's total energy use.

Figure 26 *In many parts of the world where firewood is scarce, people burn animal dung for energy. This woman is preparing cow dung that will be dried and used as fuel.*

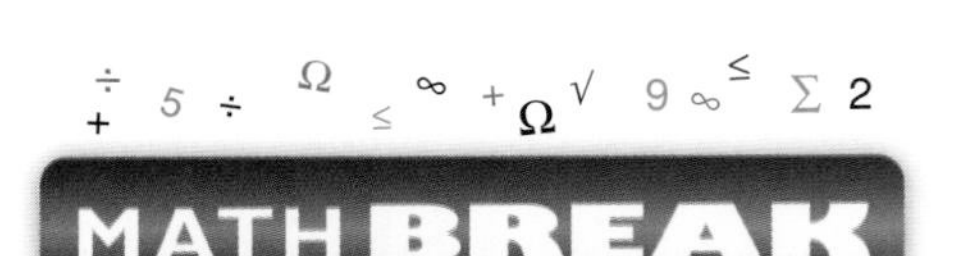

MATH BREAK

Miles per Acre

Imagine that you own a car that runs on alcohol made from corn that you grow. You drive your car about 15,000 miles in a year, and you get 240 gallons of alcohol from each acre of corn that you process. If your car gets 25 mi/gal, how many acres of corn would you have to grow to fuel your car for a year?

Gasohol Plant material can also be changed into liquid fuel. Plants containing sugar or starch, for example, can be made into alcohol. The alcohol is burned as a fuel or mixed with gasoline to make a fuel mixture called **gasohol.** An acre of corn can produce more than 1,000 L of alcohol. But in the United States we use a lot of fuel for our cars. It would take about 40 percent of the entire United States corn harvest to produce enough alcohol to make just 10 percent of the fuel we use in our cars! Biomass is obviously a renewable source of energy, but producing biomass requires land that could be used for growing food.

Deep Heat

Imagine being able to tap into the energy of the Earth. In a few places this is possible. This type of energy is called geothermal energy. **Geothermal energy** is energy produced by heat within the Earth's crust.

Geothermal Energy In some locations, rainwater penetrates porous rock near a source of magma. The heat from the magma heats the water, often turning it to steam. The steam and hot water escape through natural vents called geysers, or through wells drilled into the rock. The steam and water contain geothermal energy. Some geothermal power plants use primarily steam to generate electricity. This process is illustrated in **Figure 27.** In recent years, geothermal power plants that use primarily hot water instead of steam have become more common.

Geothermal energy can also be used as a direct source of heat. In this process, hot water and steam are used to heat a fluid that is pumped through a building in order to heat the building. Buildings in Iceland are heated in this way from the country's many geothermal sites.

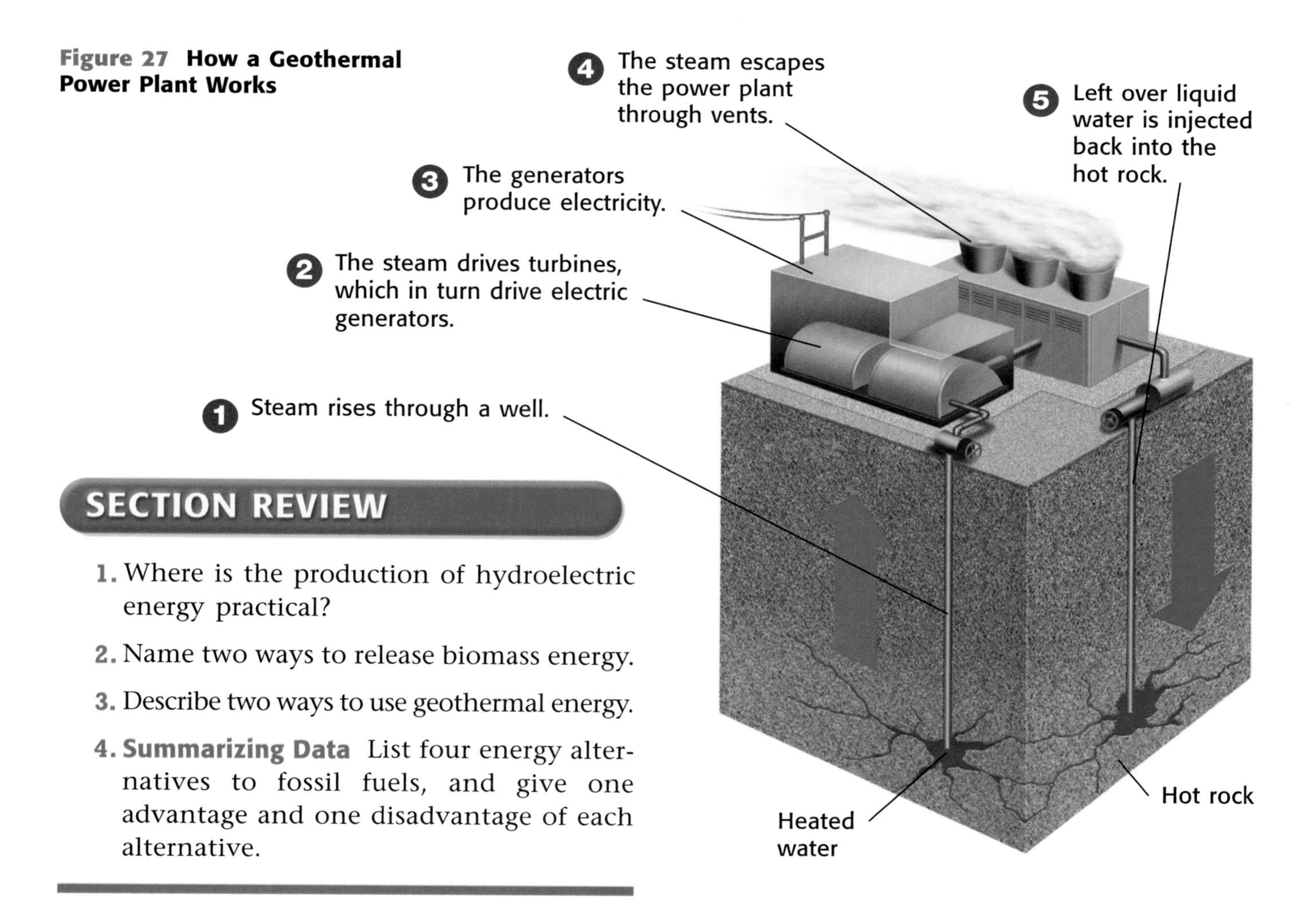

Figure 27 How a Geothermal Power Plant Works

SECTION REVIEW

1. Where is the production of hydroelectric energy practical?
2. Name two ways to release biomass energy.
3. Describe two ways to use geothermal energy.
4. **Summarizing Data** List four energy alternatives to fossil fuels, and give one advantage and one disadvantage of each alternative.

Skill Builder Lab

Make a Water Wheel

Lift Enterprises is planning to build a water wheel that will lift objects like a crane does. City planners feel that this would make very good use of the energy supplied by the river that flows through town. Development of the water wheel is in the early stages. The president of the company has asked you to modify the basic water-wheel design so that the final product will lift objects more quickly.

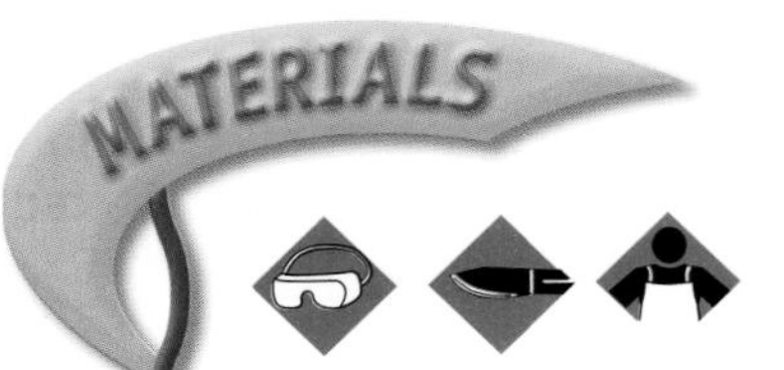

- index card
- metric ruler
- scissors
- safety razor (for teacher)
- large plastic milk jug
- permanent marker
- 5 thumbtacks
- cork
- glue
- 2 wooden skewers
- hole punch
- modeling clay
- transparent tape
- 20 cm of thread
- coin
- 2 L bottle filled with water
- watch or clock that indicates seconds

Ask a Question

1. What factors influence the rate at which a water wheel lifts a weight?

Form a Hypothesis

2. In your ScienceLog, change the question above into a statement, giving your "best guess" as to what factors will have the greatest effect on your water wheel.

Build a Model

3. Measure and mark a 5 cm × 5 cm square on an index card. Cut the square out of the card.
4. Fold the square in half to form a triangle.
5. Measure and mark a line 8 cm from the bottom of the plastic jug. Use scissors to cut along this line. (Your teacher may need to use a safety razor to start this cut for you.) Keep both sections.
6. Use the permanent marker to trace four triangles onto the flat parts of the top section of the plastic jug. Use the paper triangle you made in step 4 as a template. Cut the triangles out of the plastic to form four fins.

7. Use a thumbtack to attach one corner of each plastic fin to the round edge of the cork, as shown on page 116. Make sure the fins are equally spaced around the cork.

8. Place a drop of glue on one end of each skewer. Insert the first skewer into one of the holes in the end of the cork. Insert the second skewer into the hole in the other end.

9. Use a hole punch to carefully punch two holes in the bottom section of the plastic jug. Punch each hole 1 cm from the top edge of the jug, directly across from one another.

10. Carefully push the skewers through the holes, and suspend the cork in the center of the jug. Attach a small ball of clay to the end of each skewer. The balls should be the same size.

11. Tape one end of the thread to one skewer on the outside of the jug, next to the clay ball. Wrap the thread around the clay ball three times. (As the water wheel turns, the thread should continue to wrap around the clay. The other ball of clay balances the weight and helps to keep the water wheel turning smoothly.)

12. Tape the free end of the thread to a coin. Wrap the thread around the coin once, and tape it again.

Test the Hypothesis

13. Slowly and carefully pour water from the 2 L bottle onto the fins so that the water wheel spins. What happens to the coin? Record your observations in your ScienceLog.

14. Lower the coin back to the starting position. Add more clay to the skewer to increase the diameter of the wheel. Repeat step 13. Did the coin rise faster or slower this time?

15. Lower the coin back to the starting position. Modify the shape of the clay, and repeat step 13. Does the shape of the clay affect how quickly the coin rises? Explain your answer.

16. What happens if you remove two of the fins from opposite sides? What happens if you add more fins? Modify your water wheel to find out.

17. Experiment with another fin shape. How does a different fin shape affect how quickly the coin rises?

Analyze the Results

18. What factors influence how quickly you can lift the coin?

Draw Conclusions

19. What recommendations would you make to Lift Enterprises to improve its water wheel?

Chapter Highlights

SECTION 1

Vocabulary

natural resource *(p. 98)*
renewable resource *(p. 99)*
nonrenewable resource *(p. 99)*
recycling *(p. 100)*

Section Notes

- Natural resources include everything that is not made by humans and that can be used by organisms.
- Renewable resources, like trees and water, can be replaced in a relatively short period of time.
- Nonrenewable resources cannot be replaced, or they take a very long time to replace.
- Recycling reduces the amount of natural resources that must be obtained from the Earth.

SECTION 2

Vocabulary

energy resource *(p. 101)*
fossil fuel *(p. 101)*
petroleum *(p. 101)*
natural gas *(p. 102)*
coal *(p. 102)*
strip mining *(p. 105)*
acid precipitation *(p. 106)*
smog *(p. 107)*

Section Notes

- Fossil fuels, including petroleum, natural gas, and coal, form from the buried remains of once-living organisms.
- Petroleum and natural gas form mainly from the remains of microscopic sea life.
- Coal forms from decayed swamp plants and varies in quality based on its percentage of carbon.
- Petroleum and natural gas are obtained through drilling, while coal is obtained through mining.
- Obtaining and using fossil fuels can cause many environmental problems, including acid precipitation, water pollution, and smog.

Skills Check

Math Concepts

THE CARBON CONTENT OF COAL Turn back to page 104 to study the process of coal formation. Notice that at each stage, 10% more of the organic material becomes carbon. To calculate the percentage of carbon present at the next stage, just add 10%, or 0.10. For example:

peat → lignite
60% → 70%
$0.60 + 0.10 = 0.70$, or 70%

Visual Understanding

NO DIRECT CONTACT Take another look at Figure 21 on page 111. It is important to realize that the heated liquid inside the solar collector's tubes never comes in direct contact with the water in the tank. Cold water enters the tank, receives energy from the hot, coiled tube, and leaves the tank when someone turns on the hot-water tap.

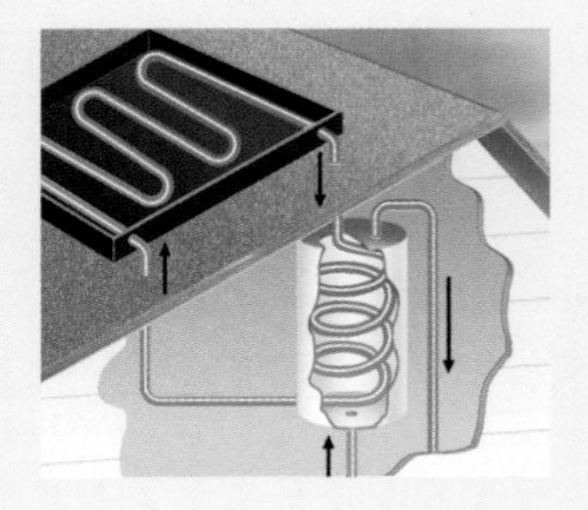

SECTION 3

Vocabulary

nuclear energy *(p. 108)*
solar energy *(p. 109)*
wind energy *(p. 112)*
hydroelectric energy *(p. 113)*
biomass *(p. 114)*
gasohol *(p. 114)*
geothermal energy *(p. 115)*

Section Notes

- Nuclear energy is most often produced by fission.
- Radioactive wastes and the threat of overheating in nuclear power plants are among the major problems associated with using nuclear energy.
- Solar energy can be converted to electricity by using solar cells.
- Solar energy can be used for direct heating by using solar collectors.
- Solar energy can be converted to electricity on both a small and large scale.
- Although harnessing wind energy is practical only in certain areas, the process produces no air pollutants, and land on wind farms can be used for more than one purpose.
- Hydroelectric energy is inexpensive, renewable, and produces little pollution. However, hydroelectric dams can damage wildlife habitats, create erosion problems, and decrease water quality.
- Plant material and animal dung that contains plant material can be burned to release energy.
- Some plant material can be converted to alcohol. This alcohol can be mixed with gasoline to make a fuel mixture called gasohol.
- Geothermal energy can be harnessed from hot, liquid water and steam that escape through natural vents or through wells drilled into the Earth's crust. This energy can be used for direct heating or can be converted to electricity.

Labs

Power of the Sun *(p. 136)*

internet**connect**

GO TO: go.hrw.com

Visit the **HRW** Web site for a variety of learning tools related to this chapter. Just type in the keyword:

KEYWORD: HSTENR

GO TO: www.scilinks.org

Visit the **National Science Teachers Association** on-line Web site for Internet resources related to this chapter. Just type in the ***sci*LINKS** number for more information about the topic:

TOPIC: Natural Resources — ***sci*LINKS NUMBER:** HSTE105
TOPIC: Renewable Resources — ***sci*LINKS NUMBER:** HSTE110
TOPIC: Nonrenewable Resources — ***sci*LINKS NUMBER:** HSTE115
TOPIC: Fossil Fuels — ***sci*LINKS NUMBER:** HSTE120
TOPIC: Nuclear Energy — ***sci*LINKS NUMBER:** HSTE122

Chapter Review

USING VOCABULARY

For each pair of terms, explain the difference in their meanings.

1. natural resource/energy resource
2. acid precipitation/smog
3. biomass/gasohol
4. hydroelectric energy/ geothermal energy

UNDERSTANDING CONCEPTS

Multiple Choice

5. Of the following, the one that is a renewable resource is
   a. coal.
   b. trees.
   c. oil.
   d. natural gas.

6. All of the following are separated from petroleum except
   a. jet fuel.
   b. lignite.
   c. kerosene.
   d. fuel oil.

7. Which of the following is a component of natural gas?
   a. gasohol
   b. methane
   c. kerosene
   d. gasoline

8. Peat, lignite, and anthracite are all stages in the formation of
   a. petroleum.
   b. natural gas.
   c. coal.
   d. gasohol.

9. Which of the following factors contribute to smog problems?
   a. high numbers of automobiles
   b. lots of sunlight
   c. mountains surrounding urban areas
   d. all of the above

10. Which of the following resources produces the least pollution?
    a. solar energy
    b. natural gas
    c. nuclear energy
    d. petroleum

11. Nuclear power plants use a process called ___?___ to produce energy.
    a. fission
    b. fusion
    c. fractionation
    d. None of the above

12. A solar-powered calculator uses
    a. solar collectors.
    b. solar panels.
    c. solar mirrors.
    d. solar cells.

13. Which of the following is a problem with using wind energy?
    a. air pollution
    b. amount of land required for wind turbines
    c. limited locations for wind farms
    d. none of the above

14. Dung is a type of
    a. geothermal energy.
    b. gasohol.
    c. biomass.
    d. None of the above

Short Answer

15. Because renewable resources can be replaced, why do we need to conserve them?
16. How does acid precipitation form?
17. If sunlight is free, why is electricity from solar cells expensive?

Concept Mapping

18. Use the following terms to create a concept map: fossil fuels, wind energy, energy resources, biomass, renewable resources, solar energy, nonrenewable resources, natural gas, gasohol, coal, oil.

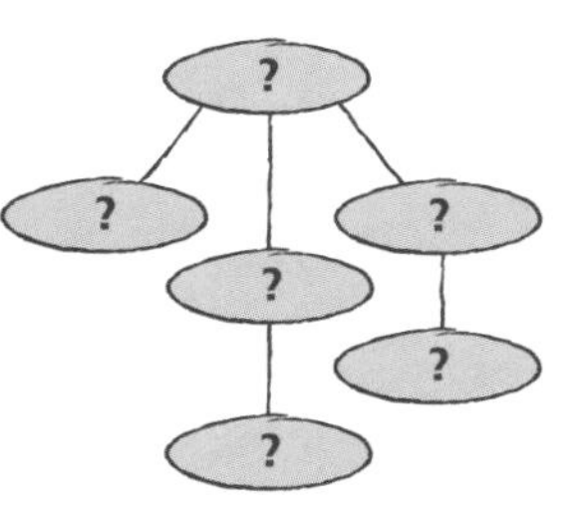

CRITICAL THINKING AND PROBLEM SOLVING

Write one or two sentences to answer the following questions:

19. How would your life be different if all fossil fuels suddenly disappeared?

20. Are fossil fuels really nonrenewable? Explain.

21. What solutions are there for the problems associated with nuclear waste?

22. How could the problems associated with the dams in Washington and local fish populations be solved?

23. What limits might there be on the productivity of a geothermal power plant?

MATH IN SCIENCE

24. Imagine that you are designing a solar car. If you mount solar cells on the underside of the car as well as on the top in direct sunlight, and it takes five times as many cells underneath to generate the same amount of electricity generated by the cells on top, what percentage of the sunlight is reflected back off the pavement?

INTERPRETING GRAPHICS

The chart below shows how various energy resources meet the world's energy needs. Use the chart to answer the following questions:

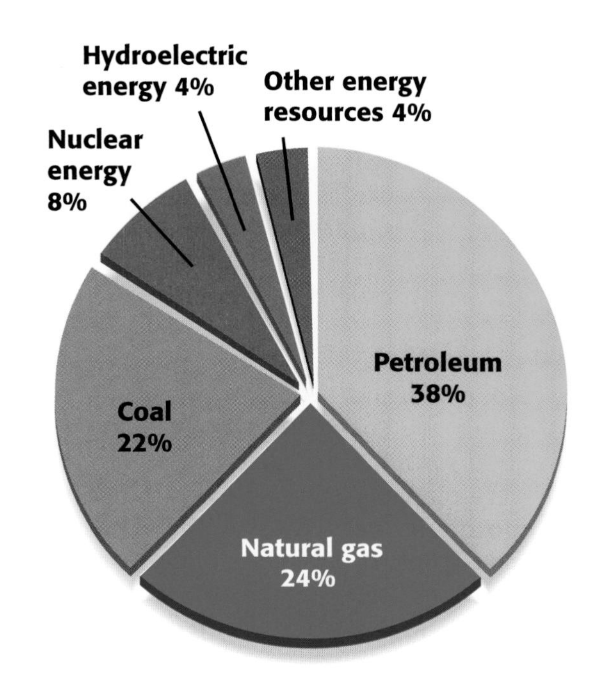

25. What percentage of the world's total energy needs is met by coal? by natural gas? by hydroelectric energy?

26. What percentage of the world's total energy needs is met by fossil fuels?

27. How much more of the world's total energy needs is met by petroleum than by natural gas?

Take a minute to review your answers to the Pre-Reading Questions found at the bottom of page 96. Have your answers changed? If necessary, revise your answers based on what you have learned since you began this chapter.

EYE ON THE ENVIRONMENT

Sitting on Your Trash

Did you know that the average person creates about 2 kg of waste every day? About 7 percent of this waste is composed of plastic products that can be recycled. Instead of adding to the landfill problem, why not recycle your plastic trash so you can sit on it? Well you can, you know! Today plastic is recycled into products like picnic tables, park benches, and even high-chairs! But how on Earth does the plastic you throw away become a park bench?

Sort It Out

Once collected and taken to a recycling center, plastic must be sorted. This process involves the coded symbols that are printed on every recyclable plastic product we use. Each product falls into one of two types of plastic—*polyethylene* or *polymer.* The plastic mainly used to make furniture includes the polyethylene plastics called *high density polyethylene,* or HDPE, and *low density polyethylene,* or LDPE. These are items such as milk jugs, detergent bottles, plastic bags, and grocery bags.

Grind It and Wash It

The recycling processes for HDPE and LDPE are fairly simple. Once it reaches the processing facility, HDPE plastic is ground into small flakes about 1 cm in diameter. In the case of LDPE plastic, which are thin films, a special grinder is used to break it down. From that point on, the recycling process is pretty much the same for LDPE and HDPE. The pieces are then washed with hot water and detergent. In this step, dirt and things like labels are removed. After the wash, the flakes are dried with blasts of hot air.

Recycle It!

Some recycling plants sell the recycled flakes. But others may reheat the flakes, change the color by adding a pigment, and then put the material in a *pelletizer.* The little pellets that result are then purchased by a company that molds the pellets into pieces of plastic lumber. This plastic lumber is used to create flowerpots, trash cans, pipes, picnic tables, park benches, toys, mats, and many other products!

From waste...

to plastic lumber...

to a park bench!

Can You Recycle It?

► The coded symbol on a plastic container tells you what type of plastic the item is made from, but it doesn't mean that you can recycle it in your area. Find out which plastics can be recycled in your state.

Oil Rush!

You may have heard of the great California gold rush. In 1849, thousands of people moved to the West hoping to strike gold. But you may not have heard about another rush that followed 10 years later. What lured people to northwestern Pennsylvania in 1859? The thrill of striking oil!

Demand for Petroleum

People began using oil as early as 3000 B.C., and oil has been a valuable substance ever since. In Mesopotamia, people used oil to waterproof their ships. The Egyptians and Chinese used oil as a medicine. It was not until the late 1700s and early 1800s that people began to use oil as a fuel. Oil was used to light homes and factories.

Petroleum Collection

But what about the oil in northwestern Pennsylvania? Did people use the oil in Pennsylvania before the rush of 1859? Native Americans were the first to dig pits to collect oil near Titusville, Pennsylvania. Early settlers used the oil as a medicine and as a fuel to light their homes. But their methods for collecting the oil were very inefficient.

The First Oil Well

In 1859, "Colonel" Edwin L. Drake came up with a better method of collecting oil from the ground. Drilling for oil! Drake hired salt-well drillers to burrow to the bedrock where oil deposits lay. But each effort was unsuccessful because water seeped into the wells, causing them to cave in. Then Drake came up with a unique idea that would make him a very wealthy man. Drake suggested that the drillers drive an iron pipe down to the bedrock 21.2 m below the surface. Then they could drill through the inner diameter of the pipe. The morning after the iron pipe was drilled, Drake woke to find that the pipe had filled with oil!

Oil City

Within 3 months, nearly 10,000 people rushed to Oil City, Pennsylvania, in search of the wealth that oil promised. Within 2 years, the small village became a bustling oil town of 50,000 people! In 1861, the first gusher well was drilled nearby, and some 3,000 barrels of oil spouted out daily. Four years later, the first oil pipeline carried crude oil a distance of 8 km.

▲ *Edwin Drake (right) and his friend Peter Wilson (left) in front of Drake Oil Well, near Titusville, Pennsylvania*

Find Out for Yourself!

► Drake's oil well was the first well used to collect oil from the ground. Research the oil wells today. How are they similar to Drake's well?

SAFETY FIRST!

Exploring, inventing, and investigating are essential to the study of science. However, these activities can also be dangerous. To make sure that your experiments and explorations are safe, you must be aware of a variety of safety guidelines.

You have probably heard of the saying, "It is better to be safe than sorry." This is particularly true in a science classroom where experiments and explorations are being performed. Being uninformed and careless can result in serious injuries. Don't take chances with your own safety or with anyone else's.

Following are important guidelines for staying safe in the science classroom. Your teacher may also have safety guidelines and tips that are specific to your classroom and laboratory. Take the time to be safe.

Safety Rules!

Start Out Right

Always get your teacher's permission before attempting any laboratory exploration. Read the procedures carefully, and pay particular attention to safety information and caution statements. If you are unsure about what a safety symbol means, look it up or ask your teacher. You cannot be too careful when it comes to safety. If an accident does occur, inform your teacher immediately, regardless of how minor you think the accident is.

If you are instructed to note the odor of a substance, wave the fumes toward your nose with your hand. Never put your nose close to the source.

Safety Symbols

All of the experiments and investigations in this book and their related worksheets include important safety symbols to alert you to particular safety concerns. Become familiar with these symbols so that when you see them, you will know what they mean and what to do. It is important that you read this entire safety section to learn about specific dangers in the laboratory.

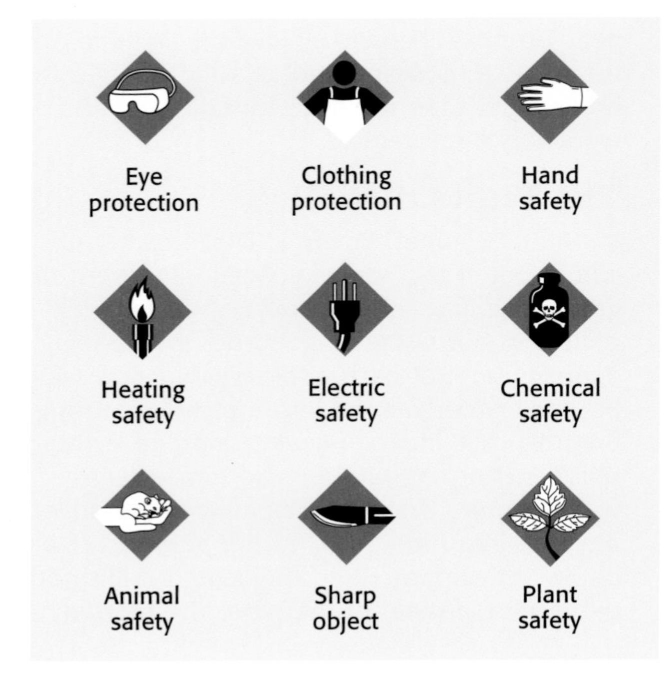

Eye Safety

Wear safety goggles when working around chemicals, acids, bases, or any type of flame or heating device. Wear safety goggles any time there is even the slightest chance that harm could come to your eyes. If any substance gets into your eyes, notify your teacher immediately, and flush your eyes with running water for at least 15 minutes. Treat any unknown chemical as if it were a dangerous chemical. Never look directly into the sun. Doing so could cause permanent blindness.

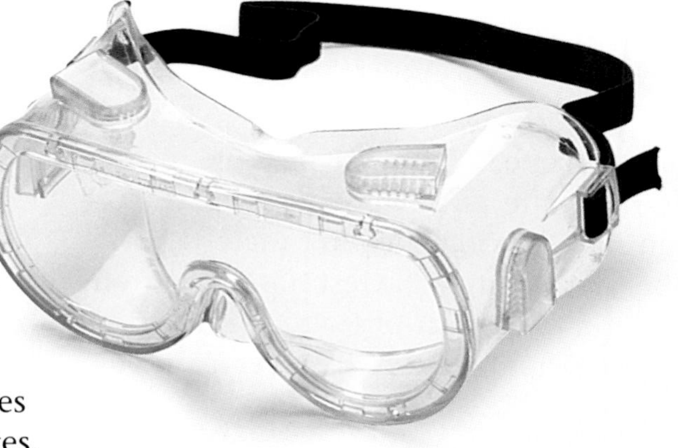

Avoid wearing contact lenses in a laboratory situation. Even if you are wearing safety goggles, chemicals can get between the contact lenses and your eyes. If your doctor requires that you wear contact lenses instead of glasses, wear eye-cup safety goggles in the lab.

Safety Equipment

Know the locations of the nearest fire alarms and any other safety equipment, such as fire blankets and eyewash fountains, as identified by your teacher, and know the procedures for using them.

Be extra careful when using any glassware. When adding a heavy object to a graduated cylinder, tilt the cylinder so the object slides slowly to the bottom.

Neatness

Keep your work area free of all unnecessary books and papers. Tie back long hair, and secure loose sleeves or other loose articles of clothing, such as ties and bows. Remove dangling jewelry. Don't wear open-toed shoes or sandals in the laboratory. Never eat, drink, or apply cosmetics in a laboratory setting. Food, drink, and cosmetics can easily become contaminated with dangerous materials.

Certain hair products (such as aerosol hair spray) are flammable and should not be worn while working near an open flame. Avoid wearing hair spray or hair gel on lab days.

Sharp/Pointed Objects

Use knives and other sharp instruments with extreme care. Never cut objects while holding them in your hands. Place objects on a suitable work surface for cutting.

Heat

Wear safety goggles when using a heating device or a flame. Whenever possible, use an electric hot plate as a heat source instead of an open flame. When heating materials in a test tube, always angle the test tube away from yourself and others. In order to avoid burns, wear heat-resistant gloves whenever instructed to do so.

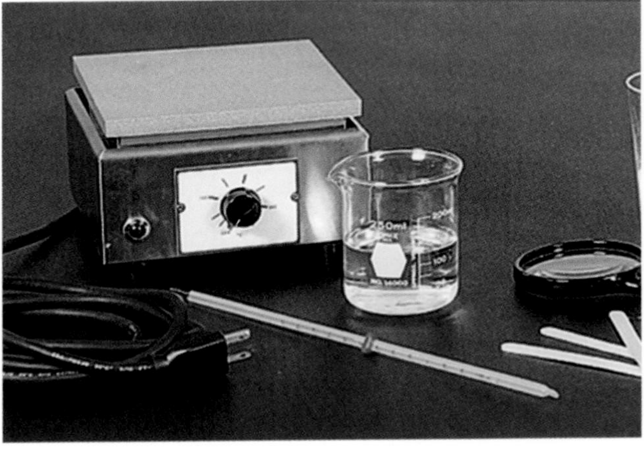

Electricity

Be careful with electrical cords. When using a microscope with a lamp, do not place the cord where it could trip someone. Do not let cords hang over a table edge in a way that could cause equipment to fall if the cord is accidentally pulled. Do not use equipment with damaged cords. Be sure your hands are dry and that the electrical equipment is in the "off" position before plugging it in. Turn off and unplug electrical equipment when you are finished.

Chemicals

Wear safety goggles when handling any potentially dangerous chemicals, acids, or bases. If a chemical is unknown, handle it as you would a dangerous chemical. Wear an apron and safety gloves when working with acids or bases or whenever you are told to do so. If a spill gets on your skin or clothing, rinse it off immediately with water for at least 5 minutes while calling to your teacher.

Never mix chemicals unless your teacher tells you to do so. Never taste, touch, or smell chemicals unless you are specifically directed to do so. Before working with a flammable liquid or gas, check for the presence of any source of flame, spark, or heat.

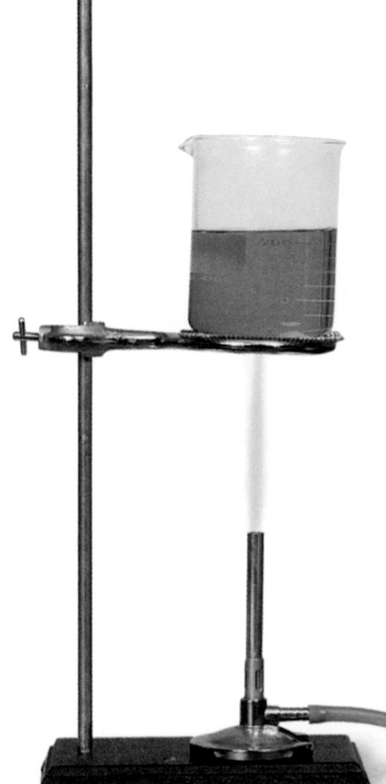

Animal Safety 

Always obtain your teacher's permission before bringing any animal into the school building. Handle animals only as your teacher directs. Always treat animals carefully and with respect. Wash your hands thoroughly after handling any animal.

Plant Safety 

Do not eat any part of a plant or plant seed used in the laboratory. Wash hands thoroughly after handling any part of a plant. When in nature, do not pick any wild plants unless your teacher instructs you to do so.

Glassware

Examine all glassware before use. Be sure that glassware is clean and free of chips and cracks. Report damaged glassware to your teacher. Glass containers used for heating should be made of heat-resistant glass.

Capturing the Wild Bean

When wildlife biologists study a group of organisms in an area, they need to know how many organisms live there. Occasionally, biologists worry that a certain organism is outgrowing the environment's carrying capacity. Other times, scientists need to know if an organism is becoming rare so steps can be taken to protect it. However, animals can be difficult to count because they can move around and hide. Because of this, biologists have developed methods to estimate the number of animals in a specific area. One of these counting methods is called the mark-recapture method.

In this activity, you will enter the territory of the wild pinto bean to get an estimation of the number of beans that live in their paper-bag habitat.

Materials

- small paper lunch bag
- pinto beans
- permanent marker
- calculator (optional)

Procedure

1. Prepare a data table in your ScienceLog like the one below.

Mark-Recapture Data Table				
Number of animals in first capture	Total number of animals in recapture	Number of marked animals in recapture	Calculated estimate of population	Actual total population
	DO NOT WRITE IN BOOK			

2. Your teacher will provide you with a paper bag containing an unknown number of beans. Carefully reach into the bag and remove a handful of beans.
3. Count the number of beans you have "captured," and record this number in your data table under "Number of animals in first capture."

4. Use the permanent marker to carefully mark each bean that you have just counted. Allow the marks to dry completely. When you are certain that all the marks are dry, place the marked beans back into the bag.
5. Gently mix the beans in the bag so the marks won't rub off. Once again, reach into the bag, "capture," and remove a handful of beans.
6. Count the number of beans in your "recapture." Record this number in your data table under "Total number of animals in recapture."
7. Count the beans in your recapture with marks from the first capture. Record this number in your data table under "Number of marked animals in recapture."
8. Calculate your estimation of the total number of beans in the bag using the following equation:

$$\frac{\text{total number of beans in recapture} \times \text{total number of beans marked}}{\text{number of marked beans in recapture}} = \text{calculated estimate of population}$$

   Enter this number in your data table under "Calculated estimate of population."
9. Replace all the beans in the bag. Then empty the bag on your work table. Be careful that no beans escape! Count each bean as you place them one at a time back into the bag. Record the number in your data table under "Actual total population."

Analysis

10. How close was your estimate to the actual number of beans?
11. If your estimate was not close to the actual number of beans, how might you change your mark-recapture procedure? If you did not recapture any marked beans, what might be the cause?

Going Further

How could you use the mark-recapture method to estimate the population of turtles in a small pond? Explain your procedure.

Nitrogen Needs

The nitrogen cycle is one of several cycles that are vital to living organisms. Without nitrogen, living organisms cannot make amino acids, the building blocks of proteins. Animals obtain nitrogen by eating plants that contain nitrogen and by eating animals that eat those plants. When animals die, decomposers return the nitrogen to the soil in the form of a chemical called ammonia.

In this activity, you will be investigating the nitrogen cycle inside a closed system to discover how decomposers return nitrogen to the soil.

Materials

- 2 pieces of filter paper
- funnel
- 50 mL beaker
- balance
- commercially prepared potting soil without fertilizer
- 25 mL graduated cylinder
- 60 mL of distilled water
- pH paper
- 1 pt (or 500 mL) jar with lid
- 5 large, dead insects from home or schoolyard
- protective gloves

Procedure

1. Fit a piece of filter paper into a funnel. Place the funnel inside a 50 mL beaker, and pour 5 g of soil into the funnel. Add 25 mL of distilled water to the soil.
2. Test the filtered water with pH paper, and record your observations in your ScienceLog.
3. Place some soil in a jar to cover the bottom about 5 cm deep. Add 10 mL of distilled water to the soil.
4. Place the dead insects in the jar, and seal the jar with the lid.
5. Check the jar each day for 5 days for an ammonia odor. (If you do not know what ammonia smells like, ask your teacher.) Record your observations in your ScienceLog. **Caution:** Your teacher will demonstrate how to check for a chemical odor by wafting. Do not put your nose in the jar and inhale!
6. On the fifth day, place a second piece of filter paper into the funnel, and place the funnel inside a 50 mL beaker. Remove about 5 g of soil from the jar, and place it in the funnel. Add 25 mL of distilled water to the soil.
7. Once again, test the filtered water with pH paper, and record your observations in your ScienceLog.

Analysis

8. What was the pH of the water in the beaker in the first trial? A pH of 7 indicates that the water is neutral. A pH below 7 indicates that the water is acidic, and a pH above 7 indicates that the water is basic.

9. What was the pH of the water in the beaker in the second trial? Explain the difference, if any, between the results of the first trial and the results of the second trial.

10. Based on the results of your pH tests, do you think ammonia is acidic or basic?

11. On which days in your investigation were you able to detect an ammonia odor? Explain what caused the odor.

12. Describe the importance of decomposers in the nitrogen cycle.

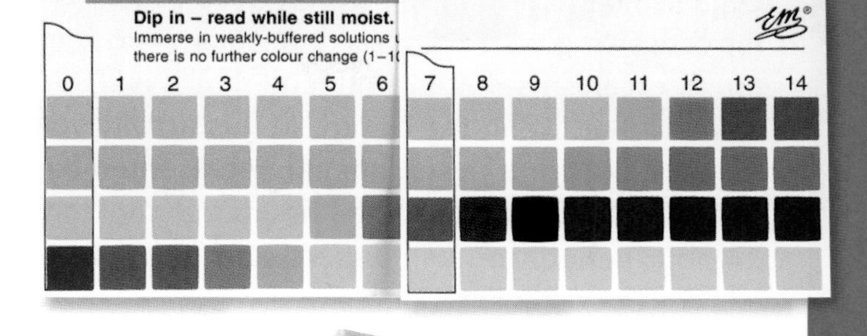

Going Further

Test ammonia's importance to plants. Fill two 12 cm flowerpots with commercially prepared potting soil and water. Be sure to use soil that has had no fertilizer added. Plant six radish seeds in each pot. Water your seeds so that the soil is constantly damp but not soaked. Keep your pots in a sunny window. You may plant other seeds of your choice, but do not use legume (bean) seeds. Research to find out why!

Use a plant fertilizer mixed according to the directions on the container to fertilize one of the pots once a week. Water the other pot once a week with tap water.

After the seedlings appear, use a metric ruler to measure the growth of the plants in both pots. Measure the plants once a week, and record your results in your ScienceLog.

Life in the Desert

Organisms that live in the desert have some very unusual methods for conserving water. Conserving water is an important function for all organisms that live on land, but it is a special challenge for animals that live in the desert. In this activity you will invent an "adaptation" for a desert animal, represented by a piece of sponge, to find out how much water the animal can conserve over a 24-hour period. You will protect your wet desert sponge so it will dry out as little as possible.

Materials

- 2 pieces of dry sponge ($8 \times 8 \times 2$ cm)
- water
- balance
- other materials as needed

Procedure

1. Plan a method for keeping your "desert animal" from drying out. Your "animal" must be in the open for at least 4 hours during the 24-hour period. Real desert animals often expose themselves to the dry desert heat in order to search for food. Write your plan in your ScienceLog. Write down your predictions about the outcome of your experiment.
2. Design data tables, if necessary, and draw them in your ScienceLog. Have your teacher approve your plan before you begin.
3. Soak two pieces of sponge in water until they begin to drip. Place each piece on a balance, and record its mass in your ScienceLog.
4. Immediately begin to protect one piece of sponge according to your plan. Place both of the pieces together in an area where they will not be disturbed. You may take your protected "animal" out for feeding as often as you want, for a total of at least 4 hours.
5. At the end of 24 hours, place each piece of sponge on the balance again, and record its mass in your ScienceLog.

Analysis

6. Describe the adaptation you used to help your "animal" survive. Was it effective? Explain.
7. What was the purpose of leaving one of the sponges unprotected? How did the water loss in each of your sponges compare?

Going Further

Conduct a class discussion about other adaptations and results. How can you relate these invented adaptations to adaptations for desert survival among real organisms?

Discovering Mini-Ecosystems

In your study of ecosystems you learned that a biome is a very large ecosystem that includes a set of smaller, related ecosystems. For example, a coniferous forest biome may include a river ecosystem, a wetland ecosystem, and a lake ecosystem. Each of those ecosystems may include several other smaller, related ecosystems. Even cities have mini-ecosystems! You may find a mini-ecosystem on a patch of sidewalk, in a puddle of rainwater, under a leaky faucet, in a shady area, or under a rock. In this activity, you will design a method for comparing two different mini-ecosystems found near your school.

Materials

- materials as needed for each investigation

Procedure

1. Examine the grounds around your school, and select two different areas you wish to investigate. Be sure to get your teacher's approval before you begin.
2. Decide what you want to learn about your mini-ecosystems. For example, you may want to know what kind of living things each area contains. You may want to list the abiotic factors of each mini-ecosystem.

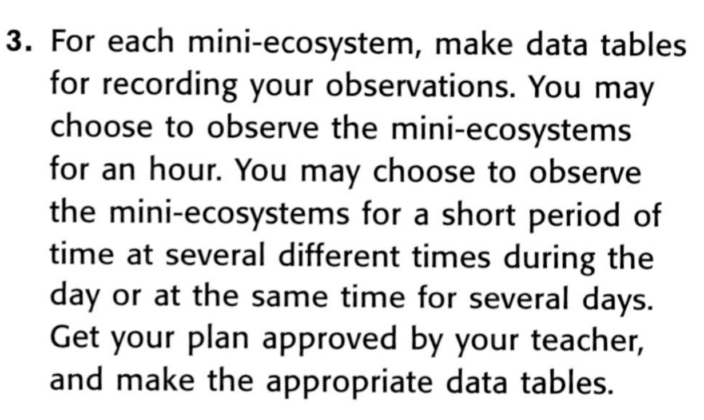

3. For each mini-ecosystem, make data tables for recording your observations. You may choose to observe the mini-ecosystems for an hour. You may choose to observe the mini-ecosystems for a short period of time at several different times during the day or at the same time for several days. Get your plan approved by your teacher, and make the appropriate data tables.

Analysis

4. What factors determine the differences between your mini-ecosystems? Identify the factors that set each mini-ecosystem apart from its surrounding area.
5. How do the populations of your mini-ecosystems compare?
6. Identify some of the adaptations that the organisms living in your two mini-ecosystems have. Describe how the adaptations help the organisms survive in their environment.
7. Write a report describing and comparing your mini-ecosystems with those of your classmates.

Using Scientific Methods

Biodiversity—What a Disturbing Thought!

Biodiversity is important for the survival of each organism in a community. Producers, consumers, and decomposers all play a cooperative role in an ecosystem.

In this activity you will investigate areas outside your school to determine which areas contain the greatest biodiversity. You will use the information you gather to determine whether a forest or an area planted with crops is more diverse.

Materials

- materials and tools necessary to carry out your investigation with your teacher's approval. Possible materials include a meterstick, binoculars, magnifying lens, twine, and forceps.

Form a Hypothesis

1. Based on your understanding of biodiversity, do you expect a forest or an area planted with crops to be more diverse?

Make a Prediction

2. Select an area that is highly disturbed (such as a mowed yard) and one that is relatively undisturbed (such as an abandoned flower bed or a vacant lot). Make a prediction about which area contains the greater biodiversity. Get your teacher's approval of your selected locations.

Conduct an Experiment

3. Design a procedure to determine which area contains the greatest biodiversity, and have your plan approved by your teacher before you begin.
4. To discover smaller organisms, measure off a square meter, set stakes at the corners, and mark the area with twine. Use a magnifying lens to observe tiny organisms. Don't worry about the scientific names. When you record your observations, refer to organisms in the following way: Ant A, Ant B, and so on. Observe each area quietly, and make note of any visits by birds or other larger organisms.

5. In your ScienceLog, create any data tables that you might need for recording your data. If you observe your areas on more than one occasion, be sure to make data tables for each observation period. Organize your data into categories that are clear and understandable.

Analyze the Results

6. Did your data support your hypothesis? Explain.
7. What factors did you consider before deciding which habitats were disturbed or undisturbed? Explain why those factors were important.
8. What problems did you find in making observations and recording data for each habitat? Describe how you solved them.
9. Describe possible errors in your investigation method. Suggest ways to improve your procedure to eliminate those errors.

Draw Conclusions

10. Do you think the biodiversity outside your school has decreased since the school was built? Why or why not?
11. Both areas shown in the photographs at right are beautiful to observe. One of them, however, is very low in biodiversity. Describe each photograph, and account for the difference in biodiversity.

Going Further

Research rain-forest biodiversity in the library or on the Internet. Find out what factors exist in the rain forest that make that biome so diverse. How might the biodiversity of a rain forest compare with that of a forest community near your school?

Prairie grasses and wildflowers

Wheat field

Power of the Sun

DISCOVERY LAB

The sun radiates energy in every direction. Like the sun, the energy radiated by a light bulb spreads out in all directions. But how much energy an object receives depends on how close that object is to the source. As you move farther from the source, the amount of energy you receive decreases. For example, if you measure the amount of energy that reaches you from a light and then move three times farther away, you will discover that nine times less energy will reach you at your second position. Energy from the sun travels as light energy. When light energy is absorbed by an object it is converted into thermal energy. *Power* is the rate at which one form of energy is converted to another, and it is measured in *watts.* Because power is related to distance, nearby objects can be used to measure the power of far-away objects. In this lab you will calculate the power of the sun using an ordinary 100-watt light bulb.

Materials

- protective gloves
- aluminum strip, 2 × 8 cm
- pencil
- black permanent marker
- Celsius thermometer
- mason jar, cap, and lid with hole in center
- modeling clay
- desk lamp with a 100 W bulb and removable shade
- metric ruler
- watch or clock that indicates seconds
- scientific calculator

Procedure

1. Gently shape the piece of aluminum around a pencil so that it holds on in the middle and has two wings, one on either side of the pencil.
2. Bend the wings outward so that they can catch as much sunlight as possible.
3. Use the marker to color both wings on one side of the aluminum strip black.
4. Remove the pencil and place the aluminum snugly around the thermometer near the bulb.
   Caution: Do not press too hard—you do not want to break the thermometer! Wear protective gloves when working with the thermometer and the aluminum.
5. Carefully slide the top of the thermometer through the hole in the lid. Place the lid on the jar so that the thermometer bulb is inside the jar, and screw down the cap.
6. Secure the thermometer to the jar lid by molding clay around the thermometer on the outside of the lid. The aluminum wings should be in the center of the jar.
7. Read the temperature on the thermometer. Record this as room temperature.
8. Place the jar on a windowsill in the sunlight. Turn the jar so that the black wings are angled toward the sun.
9. Watch the thermometer until the temperature reading stops rising. Record the temperature in your ScienceLog.
10. Remove the jar from direct sunlight, and allow it to return to room temperature.

11. Remove any shade or reflector from the lamp. Place the lamp at one end of a table.

12. Place the jar about 30 cm from the lamp. Turn the jar so that the wings are angled toward the lamp.

13. Turn on the lamp, and wait about 1 minute.

14. Move the jar a few centimeters toward the lamp until the temperature reading starts to rise. When the temperature stops rising, compare it with the reading you took in step 9.

15. Repeat step 14 until the temperature matches the temperature you recorded in step 9.

16. If the temperature reading rises too high, move the jar away from the lamp and allow it to cool. Once the reading has dropped to at least 5°C below the temperature you recorded in step 9, you may begin again at step 12.

17. When the temperature in the jar matches the temperature you recorded in step 9, record the distance between the center of the light bulb and the thermometer bulb in your ScienceLog.

Analysis

18. The thermometer measured the same amount of energy absorbed by the jar at the distance you measured to the lamp. In other words, your jar absorbed as much energy from the sun at a distance of 150 million kilometers as it did from the 100 W light bulb at the distance you recorded in step 17.

19. Use the following formula to calculate the power of the sun (be sure to show your work):

$$\frac{\text{power of the sun}}{(\text{distance to the sun})^2} = \frac{\text{power of the lamp}}{(\text{distance to the lamp})^2}$$

Hint: $(\text{distance})^2$ means that you multiply the distance by itself. If you found that the lamp was 5 cm away from the jar, for example, the $(\text{distance})^2$ would be 25.

Hint: Convert 150,000,000 km to 15,000,000,000,000 cm.

20. Review the discussion of scientific notation in the Math Refresher found in the Appendix at the back of this book. You will need to understand this technique for writing large numbers in order to compare your calculation with the actual figure. For practice, convert the distance to the sun given in step 19 to scientific notation.

$15{,}000{,}000{,}000{,}000 \text{ cm} = 1.5 \times 10^{?} \text{ cm}$

21. The sun emits 3.7×10^{26} W of power. Compare your answer in step 19 with this value. Was this a good way to calculate the power of the sun? Explain.

Concept Mapping: A Way to Bring Ideas Together

What Is a Concept Map?

Have you ever tried to tell someone about a book or a chapter you've just read and found that you can remember only a few isolated words and ideas? Or maybe you've memorized facts for a test and then weeks later discovered you're not even sure what topics those facts covered.

In both cases, you may have understood the ideas or concepts by themselves but not in relation to one another. If you could somehow link the ideas together, you would probably understand them better and remember them longer. This is something a concept map can help you do. A concept map is a way to see how ideas or concepts fit together. It can help you see the "big picture."

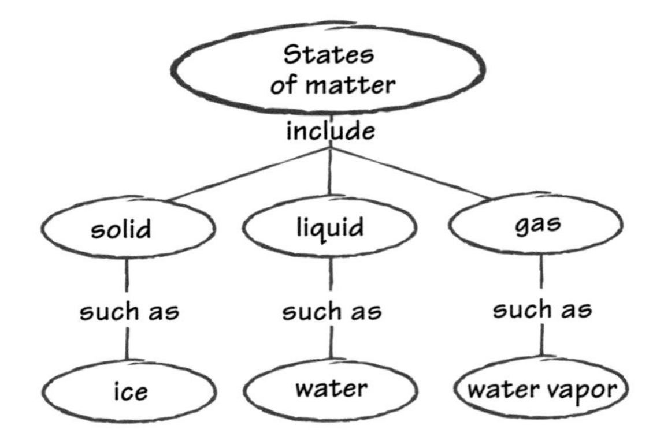

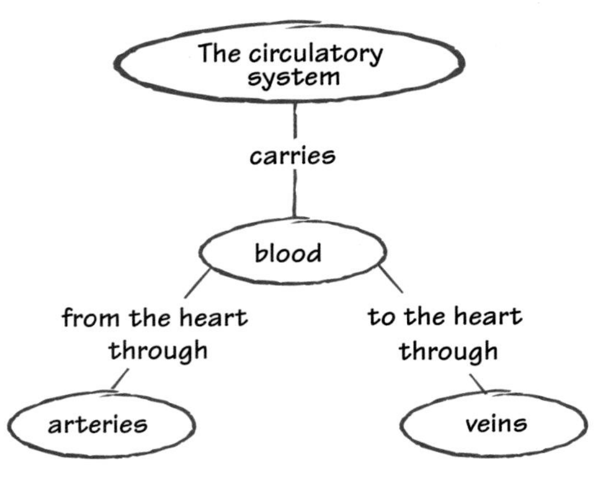

How to Make a Concept Map

1 Make a list of the main ideas or concepts.

It might help to write each concept on its own slip of paper. This will make it easier to rearrange the concepts as many times as necessary to make sense of how the concepts are connected. After you've made a few concept maps this way, you can go directly from writing your list to actually making the map.

2 Arrange the concepts in order from the most general to the most specific.

Put the most general concept at the top and circle it. Ask yourself, "How does this concept relate to the remaining concepts?" As you see the relationships, arrange the concepts in order from general to specific.

3 Connect the related concepts with lines.

4 On each line, write an action word or short phrase that shows how the concepts are related.

Look at the concept maps on this page, and then see if you can make one for the following terms:

plants, water, photosynthesis, carbon dioxide, sun's energy

One possible answer is provided at right, but don't look at it until you try the concept map yourself.

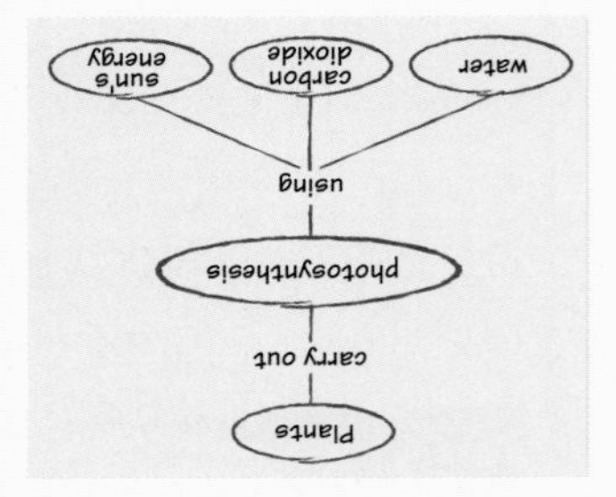

SI Measurement

The International System of Units, or SI, is the standard system of measurement used by many scientists. Using the same standards of measurement makes it easier for scientists to communicate with one another.

SI works by combining prefixes and base units. Each base unit can be used with different prefixes to define smaller and larger quantities. The table below lists common SI prefixes.

APPENDIX

SI Prefixes			
Prefix	**Abbreviation**	**Factor**	**Example**
kilo-	k	1,000	kilogram, 1 kg = 1,000 g
hecto-	h	100	hectoliter, 1 hL = 100 L
deka-	da	10	dekameter, 1 dam = 10 m
		1	meter, liter
deci-	d	0.1	decigram, 1 dg = 0.1 g
centi-	c	0.01	centimeter, 1 cm = 0.01 m
milli-	m	0.001	milliliter, 1 mL = 0.001 L
micro-	µ	0.000 001	micrometer, 1 µm = 0.000 001 m

SI Conversion Table		
SI units	**From SI to English**	**From English to SI**
Length		
kilometer (km) = 1,000 m	1 km = 0.621 mi	1 mi = 1.609 km
meter (m) = 100 cm	1 m = 3.281 ft	1 ft = 0.305 m
centimeter (cm) = 0.01 m	1 cm = 0.394 in.	1 in. = 2.540 cm
millimeter (mm) = 0.001 m	1 mm = 0.039 in.	
micrometer (µm) = 0.000 001 m		
nanometer (nm) = 0.000 000 001 m		
Area		
square kilometer (km^2) = 100 hectares	1 km^2 = 0.386 mi^2	1 mi^2 = 2.590 km^2
hectare (ha) = 10,000 m^2	1 ha = 2.471 acres	1 acre = 0.405 ha
square meter (m^2) = 10,000 cm^2	1 m^2 = 10.765 ft^2	1 ft^2 = 0.093 m^2
square centimeter (cm^2) = 100 mm^2	1 cm^2 = 0.155 $in.^2$	1 $in.^2$ = 6.452 cm^2
Volume		
liter (L) = 1,000 mL = 1 dm^3	1 L = 1.057 fl qt	1 fl qt = 0.946 L
milliliter (mL) = 0.001 L = 1 cm^3	1 mL = 0.034 fl oz	1 fl oz = 29.575 mL
microliter (µL) = 0.000 001 L		
Mass		
kilogram (kg) = 1,000 g	1 kg = 2.205 lb	1 lb = 0.454 kg
gram (g) = 1,000 mg	1 g = 0.035 oz	1 oz = 28.349 g
milligram (mg) = 0.001 g		
microgram (µg) = 0.000 001 g		

Temperature Scales

Temperature can be expressed using three different scales: Fahrenheit, Celsius, and Kelvin. The SI unit for temperature is the kelvin (K). Although 0 K is much colder than 0°C, a change of 1 K is equal to a change of 1°C.

Three Temperature Scales

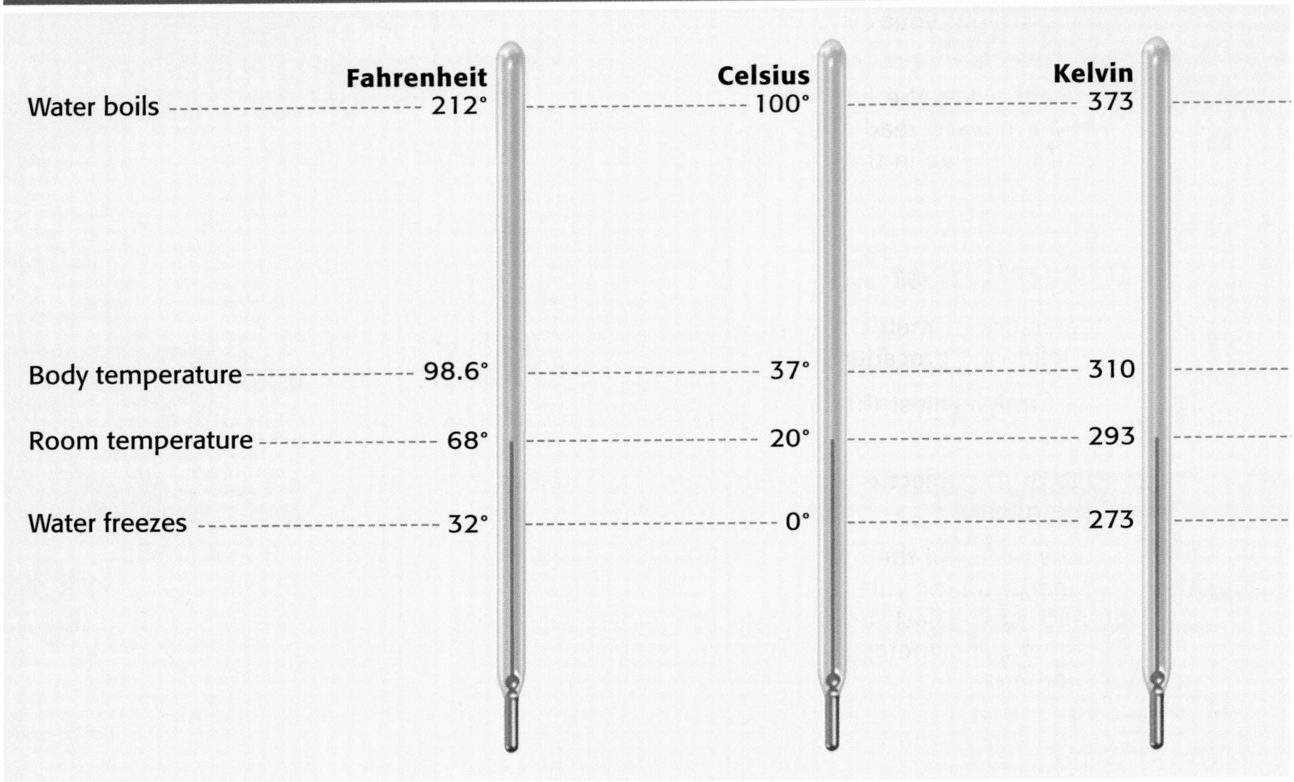

	Fahrenheit	Celsius	Kelvin
Water boils	212°	100°	373
Body temperature	98.6°	37°	310
Room temperature	68°	20°	293
Water freezes	32°	0°	273

Temperature Conversions Table

To convert	Use this equation:	Example
Celsius to Fahrenheit °C ⟶ °F	$°F = \left(\frac{9}{5} \times °C\right) + 32$	Convert 45°C to °F. $°F = \left(\frac{9}{5} \times 45°C\right) + 32 = 113°F$
Fahrenheit to Celsius °F ⟶ °C	$°C = \frac{5}{9} \times (°F - 32)$	Convert 68°F to °C. $°C = \frac{5}{9} \times (68°F - 32) = 20°C$
Celsius to Kelvin °C ⟶ K	$K = °C + 273$	Convert 45°C to K. $K = 45°C + 273 = 318\ K$
Kelvin to Celsius K ⟶ °C	$°C = K - 273$	Convert 32 K to °C. $°C = 32\ K - 273 = -241°C$

Measuring Skills

Using a Graduated Cylinder

When using a graduated cylinder to measure volume, keep the following procedures in mind:

1. Make sure the cylinder is on a flat, level surface.
2. Move your head so that your eye is level with the surface of the liquid.
3. Read the mark closest to the liquid level. On glass graduated cylinders, read the mark closest to the center of the curve in the liquid's surface.

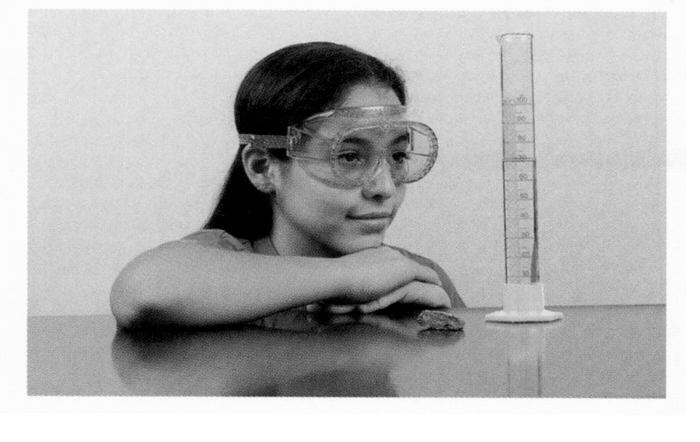

Using a Meterstick or Metric Ruler

When using a meterstick or metric ruler to measure length, keep the following procedures in mind:

1. Place the ruler firmly against the object you are measuring.
2. Align one edge of the object exactly with the zero end of the ruler.
3. Look at the other edge of the object to see which of the marks on the ruler is closest to that edge. **Note:** Each small slash between the centimeters represents a millimeter, which is one-tenth of a centimeter.

Using a Triple-Beam Balance

When using a triple-beam balance to measure mass, keep the following procedures in mind:

1. Make sure the balance is on a level surface.
2. Place all of the countermasses at zero. Adjust the balancing knob until the pointer rests at zero.
3. Place the object you wish to measure on the pan. **Caution:** Do not place hot objects or chemicals directly on the balance pan.
4. Move the largest countermass along the beam to the right until it is at the last notch that does not tip the balance. Follow the same procedure with the next-largest countermass. Then move the smallest countermass until the pointer rests at zero.
5. Add the readings from the three beams together to determine the mass of the object.
6. When determining the mass of crystals or powders, use a piece of filter paper. First find the mass of the paper. Then add the crystals or powder to the paper and remeasure. The actual mass of the crystals or powder is the total mass minus the mass of the paper. When finding the mass of liquids, first find the mass of the empty container. Then find the mass of the liquid and container together. The mass of the liquid is the total mass minus the mass of the container.

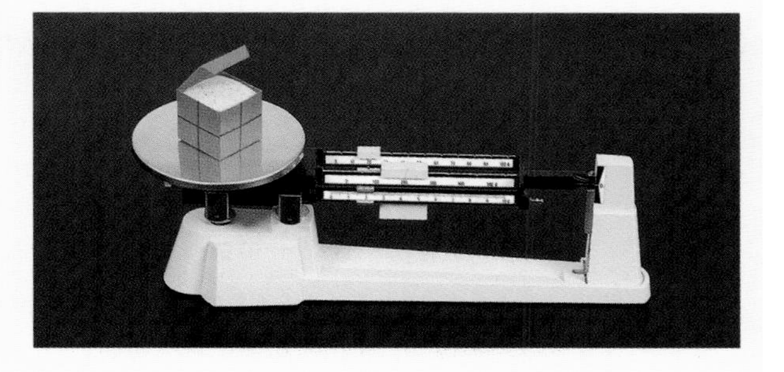

Scientific Method

The series of steps that scientists use to answer questions and solve problems is often called the **scientific method.** The scientific method is not a rigid procedure. Scientists may use all of the steps or just some of the steps of the scientific method. They may even repeat some of the steps. The goal of the scientific method is to come up with reliable answers and solutions.

Six Steps of the Scientific Method

1 **Ask a Question** Good questions come from careful **observations.** You make observations by using your senses to gather information. Sometimes you may use instruments, such as microscopes and telescopes, to extend the range of your senses. As you observe the natural world, you will discover that you have many more questions than answers. These questions drive the scientific method.

Questions beginning with *what, why, how,* and *when* are very important in focusing an investigation, and they often lead to a hypothesis. (You will learn what a hypothesis is in the next step.) Here is an example of a question that could lead to further investigation.

Question: How does acid rain affect plant growth?

Form a Hypothesis

2 **Form a Hypothesis** After you come up with a question, you need to turn the question into a **hypothesis.** A hypothesis is a clear statement of what you expect the answer to your question to be. Your hypothesis will represent your best "educated guess" based on your observations and what you already know. A good hypothesis is testable. If observations and information cannot be gathered or if an experiment cannot be designed to test your hypothesis, it is untestable, and the investigation can go no further.

Here is a hypothesis that could be formed from the question, "How does acid rain affect plant growth?"

Hypothesis: Acid rain causes plants to grow more slowly.

Notice that the hypothesis provides some specifics that lead to methods of testing. The hypothesis can also lead to predictions. A **prediction** is what you think will be the outcome of your experiment or data collection. Predictions are usually stated in an "if . . . then" format. For example, **if** meat is kept at room temperature, **then** it will spoil faster than meat kept in the refrigerator. More than one prediction can be made for a single hypothesis. Here is a sample prediction for the hypothesis that acid rain causes plants to grow more slowly.

Prediction: If a plant is watered with only acid rain (which has a pH of 4), then the plant will grow at half its normal rate.

Test the Hypothesis

3 **Test the Hypothesis** After you have formed a hypothesis and made a prediction, you should test your hypothesis. There are different ways to do this. Perhaps the most familiar way is to conduct a **controlled experiment.** A controlled experiment tests only one factor at a time. A controlled experiment has a **control group** and one or more **experimental groups.** All the factors for the control and experimental groups are the same except for one factor, which is called the **variable.** By changing only one factor, you can see the results of just that one change.

Sometimes, the nature of an investigation makes a controlled experiment impossible. For example, dinosaurs have been extinct for millions of years, and the Earth's core is surrounded by thousands of meters of rock. It would be difficult, if not impossible, to conduct controlled experiments on such things. Under such circumstances, a hypothesis may be tested by making detailed observations. Taking measurements is one way of making observations.

Analyze the Results

4 **Analyze the Results** After you have completed your experiments, made your observations, and collected your data, you must analyze all the information you have gathered. Tables and graphs are often used in this step to organize the data.

5 **Draw Conclusions** Based on the analysis of your data, you should conclude whether or not your results support your hypothesis. If your hypothesis is supported, you (or others) might want to repeat the observations or experiments to verify your results. If your hypothesis is not supported by the data, you may have to check your procedure for errors. You may even have to reject your hypothesis and make a new one. If you cannot draw a conclusion from your results, you may have to try the investigation again or carry out further observations or experiments.

Communicate Results

6 **Communicate Results** After any scientific investigation, you should report your results. By doing a written or oral report, you let others know what you have learned. They may want to repeat your investigation to see if they get the same results. Your report may even lead to another question, which in turn may lead to another investigation.

Scientific Method in Action

The scientific method is not a "straight line" of steps. It contains loops in which several steps may be repeated over and over again, while others may not be necessary. For example, sometimes scientists will find that testing one hypothesis raises new questions and new hypotheses to be tested. And sometimes, testing the hypothesis leads directly to a conclusion. Furthermore, the steps in the scientific method are not always used in the same order. Follow the steps in the diagram below, and see how many different directions the scientific method can take you.

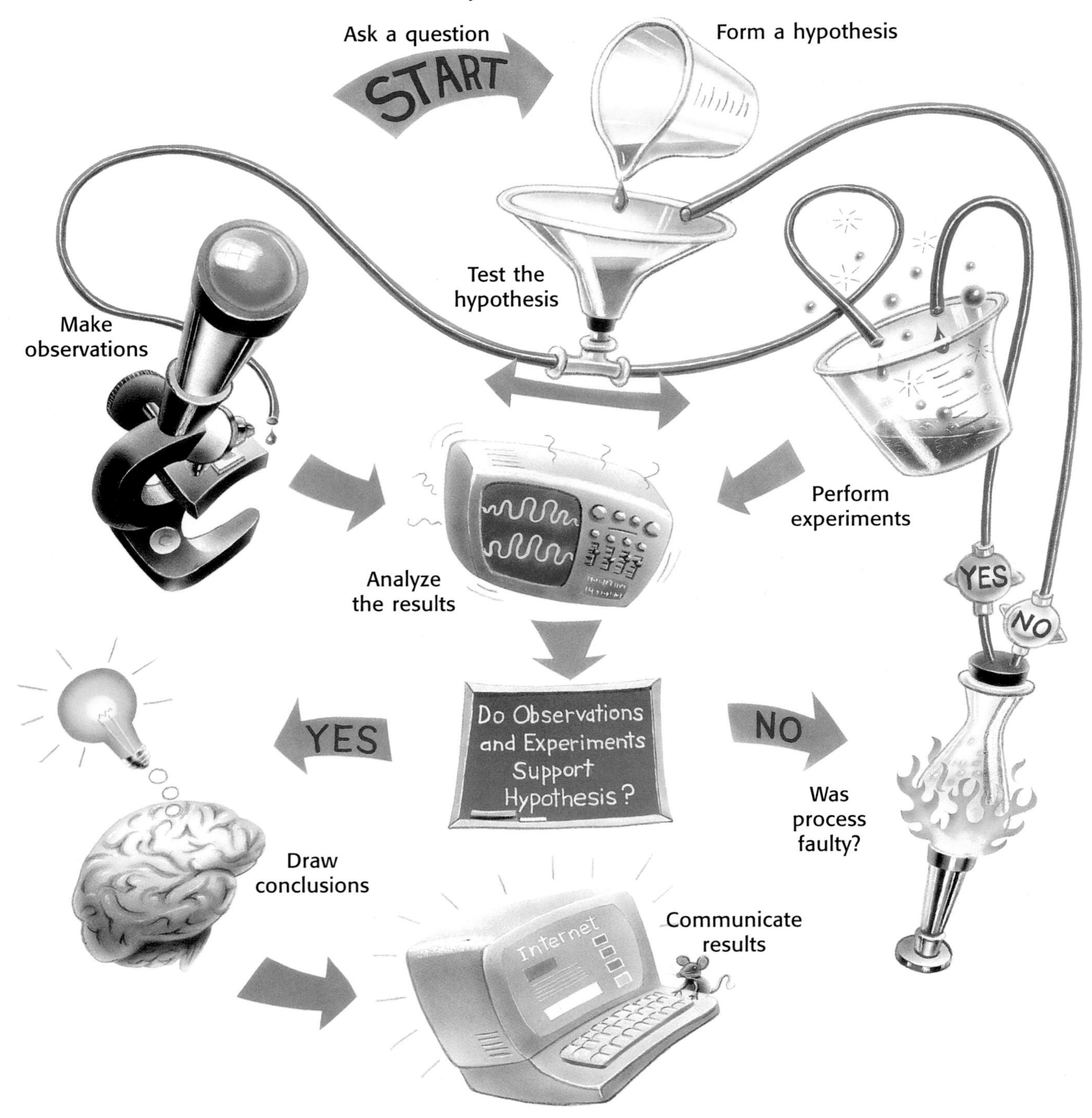

Making Charts and Graphs

Circle Graphs

A circle graph, or pie chart, shows how each group of data relates to all of the data. Each part of the circle represents a category of the data. The entire circle represents all of the data. For example, a biologist studying a hardwood forest in Wisconsin found that there were five different types of trees. The data table at right summarizes the biologist's findings.

Wisconsin Hardwood Trees	
Type of tree	**Number found**
Oak	600
Maple	750
Beech	300
Birch	1,200
Hickory	150
Total	3,000

How to Make a Circle Graph

1. In order to make a circle graph of this data, first find the percentage of each type of tree. To do this, divide the number of individual trees by the total number of trees and multiply by 100.

$$\frac{600 \text{ oak}}{3{,}000 \text{ trees}} \times 100 = 20\%$$

$$\frac{750 \text{ maple}}{3{,}000 \text{ trees}} \times 100 = 25\%$$

$$\frac{300 \text{ beech}}{3{,}000 \text{ trees}} \times 100 = 10\%$$

$$\frac{1{,}200 \text{ birch}}{3{,}000 \text{ trees}} \times 100 = 40\%$$

$$\frac{150 \text{ hickory}}{3{,}000 \text{ trees}} \times 100 = 5\%$$

2. Now determine the size of the pie shapes that make up the chart. Do this by multiplying each percentage by 360°. Remember that a circle contains 360°.

$20\% \times 360° = 72°$ $25\% \times 360° = 90°$
$10\% \times 360° = 36°$ $40\% \times 360° = 144°$
$5\% \times 360° = 18°$

3. Then check that the sum of the percentages is 100 and the sum of the degrees is 360.

$20\% + 25\% + 10\% + 40\% + 5\% = 100\%$
$72° + 90° + 36° + 144° + 18° = 360°$

4. Use a compass to draw a circle and mark its center.

5. Then use a protractor to draw angles of 72°, 90°, 36°, 144°, and 18° in the circle.

6. Finally, label each part of the graph, and choose an appropriate title.

A Community of Wisconsin Hardwood Trees

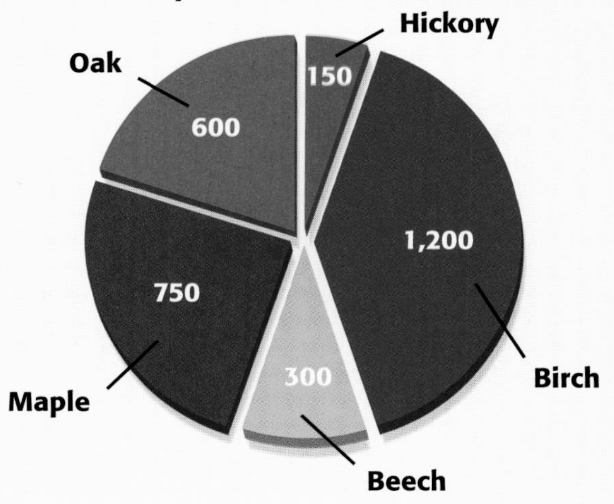

Line Graphs

Population of Appleton, 1900–2000	
Year	**Population**
1900	1,800
1920	2,500
1940	3,200
1960	3,900
1980	4,600
2000	5,300

Line graphs are most often used to demonstrate continuous change. For example, Mr. Smith's science class analyzed the population records for their hometown, Appleton, between 1900 and 2000. Examine the data at left.

Because the year and the population change, they are the *variables*. The population is determined by, or dependent on, the year. Therefore, the population is called the **dependent variable,** and the year is called the **independent variable.** Each set of data is called a **data pair.** To prepare a line graph, data pairs must first be organized in a table like the one at left.

How to Make a Line Graph

1. Place the independent variable along the horizontal (x) axis. Place the dependent variable along the vertical (y) axis.
2. Label the x-axis "Year" and the y-axis "Population." Look at your largest and smallest values for the population. Determine a scale for the y-axis that will provide enough space to show these values. You must use the same scale for the entire length of the axis. Find an appropriate scale for the x-axis too.
3. Choose reasonable starting points for each axis.
4. Plot the data pairs as accurately as possible.
5. Choose a title that accurately represents the data.

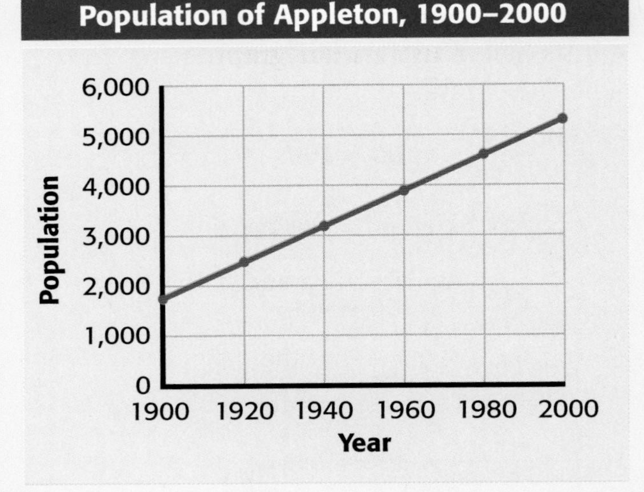

How to Determine Slope

Slope is the ratio of the change in the y-axis to the change in the x-axis, or "rise over run."

1. Choose two points on the line graph. For example, the population of Appleton in 2000 was 5,300 people. Therefore, you can define point a as (2000, 5,300). In 1900, the population was 1,800 people. Define point b as (1900, 1,800).
2. Find the change in the y-axis.
   (y at point a) − (y at point b)
   5,300 people − 1,800 people = 3,500 people
3. Find the change in the x-axis.
   (x at point a) − (x at point b)
   2000 − 1900 = 100 years
4. Calculate the slope of the graph by dividing the change in y by the change in x.

$$\text{slope} = \frac{\text{change in } y}{\text{change in } x}$$

$$\text{slope} = \frac{3{,}500 \text{ people}}{100 \text{ years}}$$

$$\text{slope} = 35 \text{ people per year}$$

In this example, the population in Appleton increased by a fixed amount each year. The graph of this data is a straight line. Therefore, the relationship is **linear.** When the graph of a set of data is not a straight line, the relationship is **nonlinear.**

Using Algebra to Determine Slope

The equation in step 4 may also be arranged to be:

$$y = kx$$

where y represents the change in the y-axis, k represents the slope, and x represents the change in the x-axis.

$$\text{slope} = \frac{\text{change in } y}{\text{change in } x}$$

$$k = \frac{y}{x}$$

$$k \times x = \frac{y \times x}{x}$$

$$kx = y$$

Bar Graphs

Bar graphs are used to demonstrate change that is not continuous. These graphs can be used to indicate trends when the data are taken over a long period of time. A meteorologist gathered the precipitation records at right for Hartford, Connecticut, for April 1–15, 1996, and used a bar graph to represent the data.

Precipitation in Hartford, Connecticut April 1–15, 1996

Date	Precipitation (cm)	Date	Precipitation (cm)
April 1	0.5	April 9	0.25
April 2	1.25	April 10	0.0
April 3	0.0	April 11	1.0
April 4	0.0	April 12	0.0
April 5	0.0	April 13	0.25
April 6	0.0	April 14	0.0
April 7	0.0	April 15	6.50
April 8	1.75		

How to Make a Bar Graph

1. Use an appropriate scale and a reasonable starting point for each axis.
2. Label the axes, and plot the data.
3. Choose a title that accurately represents the data.

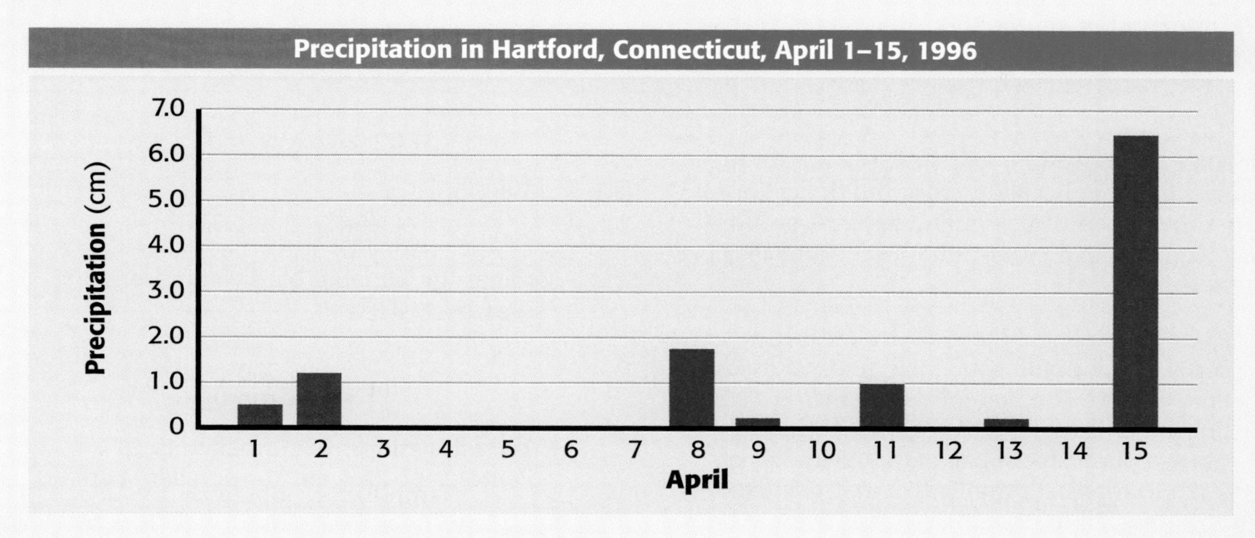

APPENDIX

Math Refresher

Science requires an understanding of many math concepts. The following pages will help you review some important math skills.

Averages

An **average,** or **mean,** simplifies a list of numbers into a single number that *approximates* their value.

Example: Find the average of the following set of numbers: 5, 4, 7, and 8.

Step 1: Find the sum.

$$5 + 4 + 7 + 8 = 24$$

Step 2: Divide the sum by the amount of numbers in your set. Because there are four numbers in this example, divide the sum by 4.

$$\frac{24}{4} = 6$$

The average, or mean, is **6.**

Ratios

A **ratio** is a comparison between numbers, and it is usually written as a fraction.

Example: Find the ratio of thermometers to students if you have 36 thermometers and 48 students in your class.

Step 1: Make the ratio.

$$\frac{\text{36 thermometers}}{\text{48 students}}$$

Step 2: Reduce the fraction to its simplest form.

$$\frac{36}{48} = \frac{36 \div 12}{48 \div 12} = \frac{3}{4}$$

The ratio of thermometers to students is **3 to 4,** or $\frac{3}{4}$. The ratio may also be written in the form 3:4.

Proportions

A **proportion** is an equation that states that two ratios are equal.

$$\frac{3}{1} = \frac{12}{4}$$

To solve a proportion, first multiply across the equal sign. This is called cross-multiplication. If you know three of the quantities in a proportion, you can use cross-multiplication to find the fourth.

Example: Imagine that you are making a scale model of the solar system for your science project. The diameter of Jupiter is 11.2 times the diameter of the Earth. If you are using a plastic-foam ball with a diameter of 2 cm to represent the Earth, what diameter does the ball representing Jupiter need to be?

$$\frac{11.2}{1} = \frac{x}{\text{2 cm}}$$

Step 1: Cross-multiply.

$$\frac{11.2}{1} \times \frac{x}{2}$$

$$11.2 \times 2 = x \times 1$$

Step 2: Multiply.

$$22.4 = x \times 1$$

Step 3: Isolate the variable by dividing both sides by 1.

$$x = \frac{22.4}{1}$$

$$x = 22.4 \text{ cm}$$

You will need to use a ball with a diameter of **22.4 cm** to represent Jupiter.

Percentages

A **percentage** is a ratio of a given number to 100.

> **Example:** What is 85 percent of 40?

Step 1: Rewrite the percentage by moving the decimal point two places to the left.

$$.85$$

Step 2: Multiply the decimal by the number you are calculating the percentage of.

$$0.85 \times 40 = 34$$

85 percent of 40 is **34.**

Decimals

To **add** or **subtract decimals,** line up the digits vertically so that the decimal points line up. Then add or subtract the columns from right to left, carrying or borrowing numbers as necessary.

> **Example:** Add the following numbers: 3.1415 and 2.96.

Step 1: Line up the digits vertically so that the decimal points line up.

$$\begin{array}{r} 3.1415 \\ +\ 2.96 \\ \hline \end{array}$$

Step 2: Add the columns from right to left, carrying when necessary.

$$\begin{array}{r} {\scriptstyle 1\ 1} \\ 3.1415 \\ +\ 2.96 \\ \hline 6.1015 \end{array}$$

The sum is **6.1015.**

Fractions

Numbers tell you how many; **fractions** tell you *how much of a whole.*

> **Example:** Your class has 24 plants. Your teacher instructs you to put 5 in a shady spot. What fraction does this represent?

Step 1: Write a fraction with the total number of parts in the whole as the denominator.

$$\frac{?}{24}$$

Step 2: Write the number of parts of the whole being represented as the numerator.

$$\frac{5}{24}$$

$\mathbf{\frac{5}{24}}$ of the plants will be in the shade.

Reducing Fractions

It is usually best to express a fraction in simplest form. This is called *reducing* a fraction.

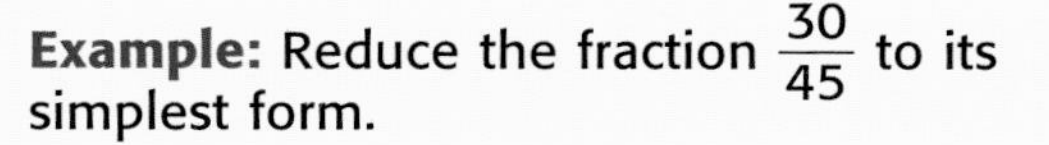

> **Example:** Reduce the fraction $\frac{30}{45}$ to its simplest form.

Step 1: Find the largest whole number that will divide evenly into both the numerator and denominator. This number is called the greatest common factor (GCF).

factors of the numerator 30: 1, 2, 3, 5, 6, 10, **15,** 30

factors of the denominator 45: 1, 3, 5, 9, **15,** 45

Step 2: Divide both the numerator and the denominator by the GCF, which in this case is 15.

$$\frac{30}{45} = \frac{30 \div 15}{45 \div 15} = \frac{2}{3}$$

$\frac{30}{45}$ reduced to its simplest form is $\mathbf{\frac{2}{3}}$**.**

Adding and Subtracting Fractions

To **add** or **subtract fractions** that have the **same denominator,** simply add or subtract the numerators.

Examples:

$$\frac{3}{5} + \frac{1}{5} = ? \text{ and } \frac{3}{4} - \frac{1}{4} = ?$$

Step 1: Add or subtract the numerators.

$$\frac{3}{5} + \frac{1}{5} = \frac{4}{} \text{ and } \frac{3}{4} - \frac{1}{4} = \frac{2}{}$$

Step 2: Write the sum or difference over the denominator.

$$\frac{3}{5} + \frac{1}{5} = \frac{4}{5} \text{ and } \frac{3}{4} - \frac{1}{4} = \frac{2}{4}$$

Step 3: If necessary, reduce the fraction to its simplest form.

$\mathbf{\frac{4}{5}}$ cannot be reduced, and $\frac{2}{4} = \mathbf{\frac{1}{2}}$.

To **add** or **subtract fractions** that have **different denominators,** first find the least common denominator (LCD).

Examples:

$$\frac{1}{2} + \frac{1}{6} = ? \text{ and } \frac{3}{4} - \frac{2}{3} = ?$$

Step 1: Write the equivalent fractions with a common denominator.

$$\frac{3}{6} + \frac{1}{6} = ? \text{ and } \frac{9}{12} - \frac{8}{12} = ?$$

Step 2: Add or subtract.

$$\frac{3}{6} + \frac{1}{6} = \frac{4}{6} \text{ and } \frac{9}{12} - \frac{8}{12} = \frac{1}{12}$$

Step 3: If necessary, reduce the fraction to its simplest form.

$\frac{4}{6} = \mathbf{\frac{2}{3}}$, and $\mathbf{\frac{1}{12}}$ cannot be reduced.

Multiplying Fractions

To **multiply fractions,** multiply the numerators and the denominators together, and then reduce the fraction to its simplest form.

Example:

$$\frac{5}{9} \times \frac{7}{10} = ?$$

Step 1: Multiply the numerators and denominators.

$$\frac{5}{9} \times \frac{7}{10} = \frac{5 \times 7}{9 \times 10} = \frac{35}{90}$$

Step 2: Reduce.

$$\frac{35}{90} = \frac{35 \div 5}{90 \div 5} = \mathbf{\frac{7}{18}}$$

Dividing Fractions

To **divide fractions,** first rewrite the divisor (the number you divide *by*) upside down. This is called the reciprocal of the divisor. Then you can multiply and reduce if necessary.

Example:

$$\frac{5}{8} \div \frac{3}{2} = ?$$

Step 1: Rewrite the divisor as its reciprocal.

$$\frac{3}{2} \rightarrow \frac{2}{3}$$

Step 2: Multiply.

$$\frac{5}{8} \times \frac{2}{3} = \frac{5 \times 2}{8 \times 3} = \frac{10}{24}$$

Step 3: Reduce.

$$\frac{10}{24} = \frac{10 \div 2}{24 \div 2} = \mathbf{\frac{5}{12}}$$

Scientific Notation

Scientific notation is a short way of representing very large and very small numbers without writing all of the place-holding zeros.

Example: Write 653,000,000 in scientific notation.

Step 1: Write the number without the place-holding zeros.

653

Step 2: Place the decimal point after the first digit.

6.53

Step 3: Find the exponent by counting the number of places that you moved the decimal point.

6.53000000

The decimal point was moved eight places to the left. Therefore, the exponent of 10 is positive 8. Remember, if the decimal point had moved to the right, the exponent would be negative.

Step 4: Write the number in scientific notation.

$\mathbf{6.53 \times 10^8}$

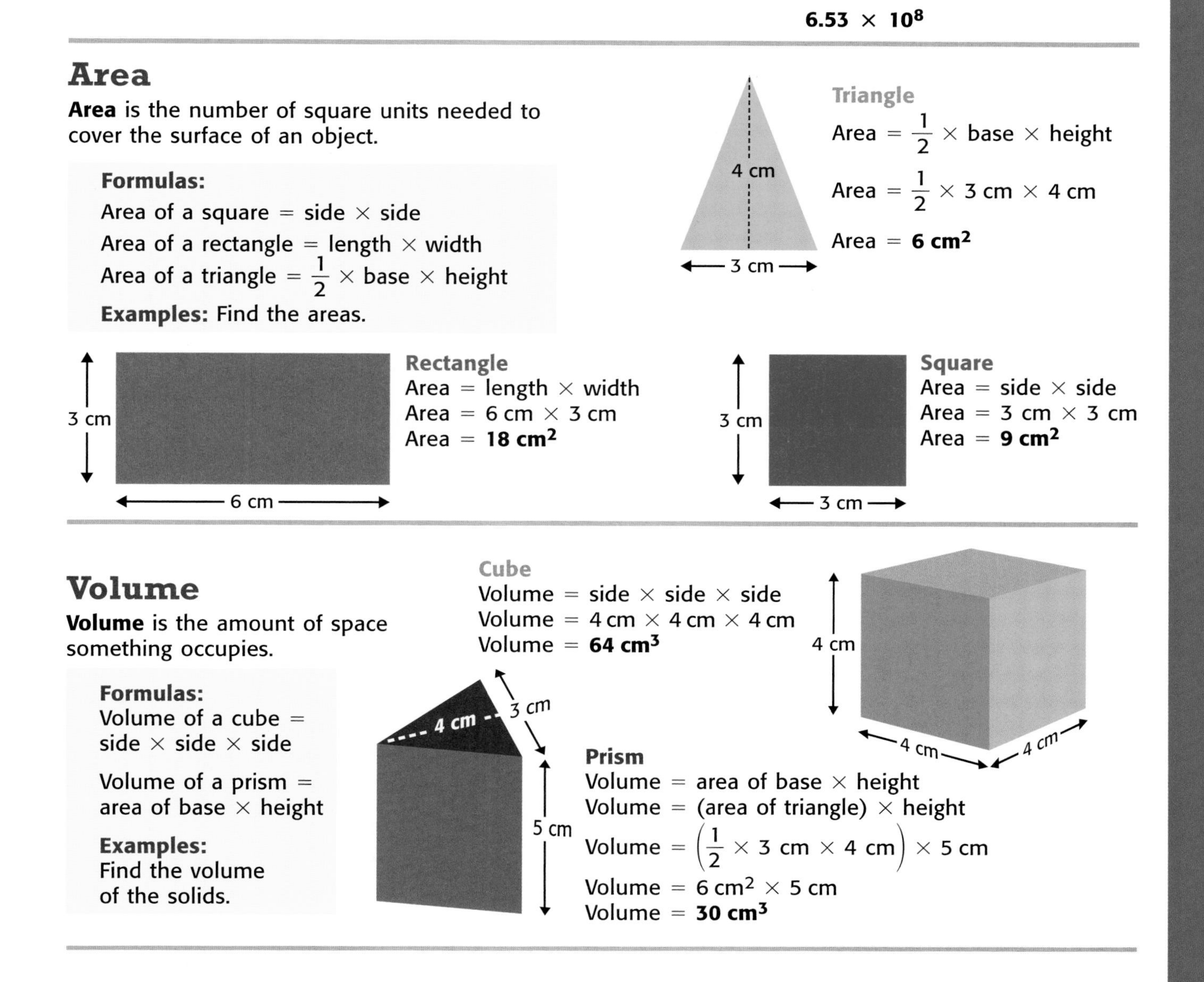

Area

Area is the number of square units needed to cover the surface of an object.

Formulas:

Area of a square = side × side

Area of a rectangle = length × width

Area of a triangle = $\frac{1}{2}$ × base × height

Examples: Find the areas.

Triangle

Area = $\frac{1}{2}$ × base × height

Area = $\frac{1}{2}$ × 3 cm × 4 cm

Area = **6 cm^2**

Rectangle

Area = length × width

Area = 6 cm × 3 cm

Area = **18 cm^2**

Square

Area = side × side

Area = 3 cm × 3 cm

Area = **9 cm^2**

Volume

Volume is the amount of space something occupies.

Formulas:

Volume of a cube = side × side × side

Volume of a prism = area of base × height

Examples:

Find the volume of the solids.

Cube

Volume = side × side × side

Volume = 4 cm × 4 cm × 4 cm

Volume = **64 cm^3**

Prism

Volume = area of base × height

Volume = (area of triangle) × height

Volume = $\left(\frac{1}{2} \times 3 \text{ cm} \times 4 \text{ cm}\right) \times 5 \text{ cm}$

Volume = 6 cm^2 × 5 cm

Volume = **30 cm^3**

Periodic Table of the Elements

Each square on the table includes an element's name, chemical symbol, atomic number, and atomic mass.

Atomic number — 6
Chemical symbol — C
Element name — Carbon
Atomic mass — 12.0

The background color indicates the type of element. Carbon is a nonmetal.

The color of the chemical symbol indicates the physical state at room temperature. Carbon is a solid.

Background
- Metals
- Metalloids
- Nonmetals

Chemical Symbol
- Solid
- Liquid
- Gas

	Group 1	Group 2	Group 3	Group 4	Group 5	Group 6	Group 7	Group 8	Group 9
Period 1	1 H Hydrogen 1.0								
Period 2	3 Li Lithium 6.9	4 Be Beryllium 9.0							
Period 3	11 Na Sodium 23.0	12 Mg Magnesium 24.3							
Period 4	19 K Potassium 39.1	20 Ca Calcium 40.1	21 Sc Scandium 45.0	22 Ti Titanium 47.9	23 V Vanadium 50.9	24 Cr Chromium 52.0	25 Mn Manganese 54.9	26 Fe Iron 55.8	27 Co Cobalt 58.9
Period 5	37 Rb Rubidium 85.5	38 Sr Strontium 87.6	39 Y Yttrium 88.9	40 Zr Zirconium 91.2	41 Nb Niobium 92.9	42 Mo Molybdenum 95.9	43 Tc Technetium (97.9)	44 Ru Ruthenium 101.1	45 Rh Rhodium 102.9
Period 6	55 Cs Cesium 132.9	56 Ba Barium 137.3	57 La Lanthanum 138.9	72 Hf Hafnium 178.5	73 Ta Tantalum 180.9	74 W Tungsten 183.8	75 Re Rhenium 186.2	76 Os Osmium 190.2	77 Ir Iridium 192.2
Period 7	87 Fr Francium (223.0)	88 Ra Radium (226.0)	89 Ac Actinium (227.0)	104 Rf Rutherfordium (261.1)	105 Db Dubnium (262.1)	106 Sg Seaborgium (263.1)	107 Bh Bohrium (262.1)	108 Hs Hassium (265)	109 Mt Meitnerium (266)

A row of elements is called a period.

A column of elements is called a group or family.

Lanthanides	58 Ce Cerium 140.1	59 Pr Praseodymium 140.9	60 Nd Neodymium 144.2	61 Pm Promethium (144.9)	62 Sm Samarium 150.4
Actinides	90 Th Thorium 232.0	91 Pa Protactinium 231.0	92 U Uranium 238.0	93 Np Neptunium (237.0)	94 Pu Plutonium 244.1

These elements are placed below the table to allow the table to be narrower.

Group 10	Group 11	Group 12	Group 13	Group 14	Group 15	Group 16	Group 17	Group 18
								2 **He** Helium 4.0
			5 **B** Boron 10.8	6 **C** Carbon 12.0	7 **N** Nitrogen 14.0	8 **O** Oxygen 16.0	9 **F** Fluorine 19.0	10 **Ne** Neon 20.2
			13 **Al** Aluminum 27.0	14 **Si** Silicon 28.1	15 **P** Phosphorus 31.0	16 **S** Sulfur 32.1	17 **Cl** Chlorine 35.5	18 **Ar** Argon 39.9
28 **Ni** Nickel 58.7	29 **Cu** Copper 63.5	30 **Zn** Zinc 65.4	31 **Ga** Gallium 69.7	32 **Ge** Germanium 72.6	33 **As** Arsenic 74.9	34 **Se** Selenium 79.0	35 **Br** Bromine 79.9	36 **Kr** Krypton 83.8
46 **Pd** Palladium 106.4	47 **Ag** Silver 107.9	48 **Cd** Cadmium 112.4	49 **In** Indium 114.8	50 **Sn** Tin 118.7	51 **Sb** Antimony 121.8	52 **Te** Tellurium 127.6	53 **I** Iodine 126.9	54 **Xe** Xenon 131.3
78 **Pt** Platinum 195.1	79 **Au** Gold 197.0	80 **Hg** Mercury 200.6	81 **Tl** Thallium 204.4	82 **Pb** Lead 207.2	83 **Bi** Bismuth 209.0	84 **Po** Polonium (209.0)	85 **At** Astatine (210.0)	86 **Rn** Radon (222.0)
110 **Uun*** Ununnilium (271)	111 **Uuu*** Unununium (272)	112 **Uub*** Ununbium (277)		114 **Uuq*** Ununquadium (285)		116 **Uuh*** Ununhexium (289)		118 **Uuo*** Ununoctium (293)

This zigzag line reminds you where the metals, nonmetals, and metalloids are.

A number in parenthesis is the mass number of the most stable form of that element.

63 **Eu** Europium 152.0	64 **Gd** Gadolinium 157.3	65 **Tb** Terbium 158.9	66 **Dy** Dysprosium 162.5	67 **Ho** Holmium 164.9	68 **Er** Erbium 167.3	69 **Tm** Thulium 168.9	70 **Yb** Ytterbium 173.0	71 **Lu** Lutetium 175.0
95 **Am** Americium (243.1)	96 **Cm** Curium (247.1)	97 **Bk** Berkelium (247.1)	98 **Cf** Californium (251.1)	99 **Es** Einsteinium (252.1)	100 **Fm** Fermium (257.1)	101 **Md** Mendelevium (258.1)	102 **No** Nobelium (259.1)	103 **Lr** Lawrencium (262.1)

**The official names and symbols for the elements greater than 109 will eventually be approved by a committee of scientists.*

APPENDIX

Physical Science Refresher

Atoms and Elements

Every object in the universe is made up of particles of some kind of matter. **Matter** is anything that takes up space and has mass. All matter is made up of elements. An **element** is a substance that cannot be separated into simpler components by ordinary chemical means. This is because each element consists of only one kind of atom. An **atom** is the smallest unit of an element that has all of the properties of that element.

Atomic Structure

Atoms are made up of small particles called subatomic particles. The three major types of subatomic particles are **electrons, protons,** and **neutrons.** Electrons have a negative electric charge, protons have a positive charge, and neutrons have no electric charge. The protons and neutrons are packed close to one another to form the **nucleus.** The protons give the nucleus a positive charge. Electrons are most likely to be found in regions around the nucleus called **electron clouds.** The negatively charged electrons are attracted to the positively charged nucleus. An atom may have several energy levels in which electrons are located.

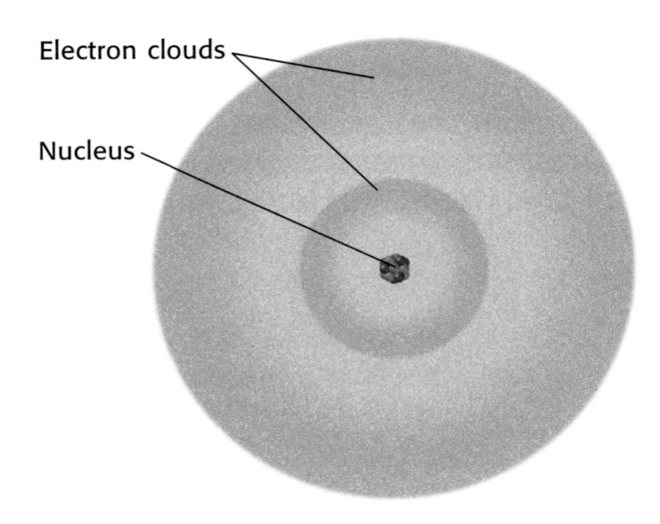

Atomic Number

To help in the identification of elements, scientists have assigned an **atomic number** to each kind of atom. The atomic number is the number of protons in the atom. Atoms with the same number of protons are all the same kind of element. In an uncharged, or electrically neutral, atom there are an equal number of protons and electrons. Therefore, the atomic number equals the number of electrons in an uncharged atom. The number of neutrons, however, can vary for a given element. Atoms of the same element that have different numbers of neutrons are called **isotopes.**

Periodic Table of the Elements

In the periodic table, the elements are arranged from left to right in order of increasing atomic number. Each element in the table is in a separate box. An atom of each element has one more electron and one more proton than an atom of the element to its left. Each horizontal row of the table is called a **period.** Changes in chemical properties of elements across a period correspond to changes in the electron arrangements of their atoms. Each vertical column of the table, known as a **group,** lists elements with similar properties. The elements in a group have similar chemical properties because their atoms have the same number of electrons in their outer energy level. For example, the elements helium, neon, argon, krypton, xenon, and radon all have similar properties and are known as the noble gases.

Molecules and Compounds

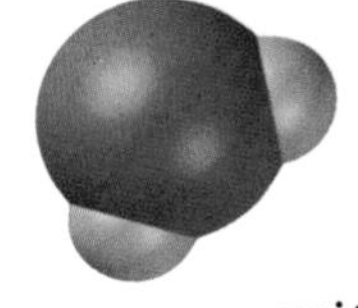

When two or more elements are joined chemically, the resulting substance is called a **compound.** A compound is a new substance with properties different from those of the elements that compose it. For example, water, H_2O, is a compound formed when hydrogen (H) and oxygen (O) combine. The smallest complete unit of a compound that has the properties of that compound is called a **molecule.** A chemical formula indicates the elements in a compound. It also indicates the relative number of atoms of each element present. The chemical formula for water is H_2O, which indicates that each water molecule consists of two atoms of hydrogen and one atom of oxygen. The subscript number is used after the symbol for an element to indicate how many atoms of that element are in a single molecule of the compound.

Acids, Bases, and pH

An ion is an atom or group of atoms that has an electric charge because it has lost or gained one or more electrons. When an acid, such as hydrochloric acid, HCl, is mixed with water, it separates into ions. An **acid** is a compound that produces hydrogen ions, H^+, in water. The hydrogen ions then combine with a water molecule to form a hydronium ion, H_3O^+. A **base,** on the other hand, is a substance that produces hydroxide ions, OH^-, in water.

To determine whether a solution is acidic or basic, scientists use pH. The **pH** is a measure of the hydronium ion concentration in a solution. The pH scale ranges from 0 to 14. The middle point, pH = 7, is neutral, neither acidic nor basic. Acids have a pH less than 7; bases have a pH greater than 7. The lower the number is, the more acidic the solution. The higher the number is, the more basic the solution.

Chemical Equations

A chemical reaction occurs when a chemical change takes place. (In a chemical change, new substances with new properties are formed.) A chemical equation is a useful way of describing a chemical reaction by means of chemical formulas. The equation indicates what substances react and what the products are. For example, when carbon and oxygen combine, they can form carbon dioxide. The equation for the reaction is as follows: $C + O_2 \rightarrow CO_2$.

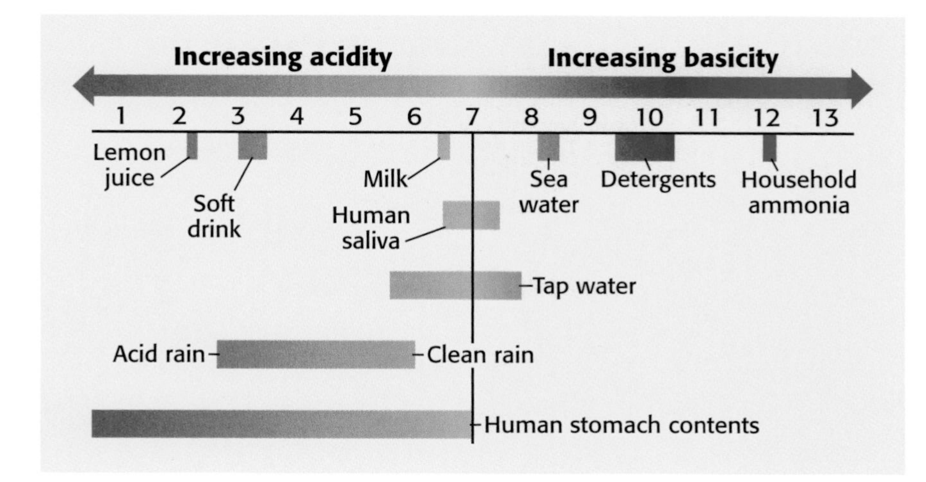

APPENDIX

APPENDIX

The Six Kingdoms

Kingdom Archaebacteria

The organisms in this kingdom are single-celled prokaryotes.

Archaebacteria		
Group	**Examples**	**Characteristics**
Methanogens	*Methanococcus*	found in soil, swamps, the digestive tract of mammals; produce methane gas; can't live in oxygen
Thermophiles	*Sulpholobus*	found in extremely hot environments; require sulphur, can't live in oxygen
Halophiles	*Halococcus*	found in environments with very high salt content, such as the Dead Sea; nearly all can live in oxygen

Kingdom Eubacteria

There are more than 4,000 named species in this kingdom of single-celled prokaryotes.

Eubacteria		
Group	**Examples**	**Characteristics**
Bacilli	*Escherichia coli*	rod-shaped; free-living, symbiotic, or parasitic; some can fix nitrogen; some cause disease
Cocci	*Streptococcus*	spherical-shaped, disease-causing; can form spores to resist unfavorable environments
Spirilla	*Treponema*	spiral-shaped; responsible for several serious illnesses, such as syphilis and Lyme disease

Kingdom Protista

The organisms in this kingdom are eukaryotes. There are single-celled and multicellular representatives.

Protists		
Group	**Examples**	**Characteristics**
Sacodines	*Amoeba*	radiolarians; single-celled consumers
Ciliates	*Paramecium*	single-celled consumers
Flagellates	*Trypanosoma*	single-celled parasites
Sporozoans	*Plasmodium*	single-celled parasites
Euglenas	*Euglena*	single-celled; photosynthesize
Diatoms	*Pinnularia*	most are single-celled; photosynthesize
Dinoflagellates	*Gymnodinium*	single-celled; some photosynthesize
Algae	*Volvox,* coral algae	4 phyla; single- or many-celled; photosynthesize
Slime molds	*Physarum*	single- or many-celled; consumers or decomposers
Water molds	powdery mildew	single- or many-celled, parasites or decomposers

Kingdom Fungi

There are single-celled and multicellular eukaryotes in this kingdom. There are four major groups of fungi.

Fungi		
Group	**Examples**	**Characteristics**
Threadlike fungi	bread mold	spherical; decomposers
Sac fungi	yeast, morels	saclike; parasites and decomposers
Club fungi	mushrooms, rusts, smuts	club-shaped; parasites and decomposers
Lichens	British soldier	symbiotic with algae

Kingdom Plantae

The organisms in this kingdom are multicellular eukaryotes. They have specialized organ systems for different life processes. They are classified in divisions instead of phyla.

Plants		
Group	**Examples**	**Characteristics**
Bryophytes	mosses, liverworts	reproduce by spores
Club mosses	*Lycopodium,* ground pine	reproduce by spores
Horsetails	rushes	reproduce by spores
Ferns	spleenworts, sensitive fern	reproduce by spores
Conifers	pines, spruces, firs	reproduce by seeds; cones
Cycads	*Zamia*	reproduce by seeds
Gnetophytes	*Welwitschia*	reproduce by seeds
Ginkgoes	*Ginkgo*	reproduce by seeds
Angiosperms	all flowering plants	reproduce by seeds; flowers

Kingdom Animalia

This kingdom contains multicellular eukaryotes. They have specialized tissues and complex organ systems.

Animals		
Group	**Examples**	**Characteristics**
Sponges	glass sponges	no symmetry or segmentation; aquatic
Cnidarians	jellyfish, coral	radial symmetry; aquatic
Flatworms	planaria, tapeworms, flukes	bilateral symmetry; organ systems
Roundworms	*Trichina,* hookworms	bilateral symmetry; organ systems
Annelids	earthworms, leeches	bilateral symmetry; organ systems
Mollusks	snails, octopuses	bilateral symmetry; organ systems
Echinoderms	sea stars, sand dollars	radial symmetry; organ systems
Arthropods	insects, spiders, lobsters	bilateral symmetry; organ systems
Chordates	fish, amphibians, reptiles, birds, mammals	bilateral symmetry; complex organ systems

Using the Microscope

Parts of the Compound Light Microscope

- The **ocular lens** magnifies the image 10×.
- The **low-power objective** magnifies the image 10×.
- The **high-power objective** magnifies the image either 40× or 43×.
- The **revolving nosepiece** holds the objectives and can be turned to change from one magnification to the other.
- The **body tube** maintains the correct distance between the ocular lens and objectives.
- The **coarse-adjustment knob** moves the body tube up and down to allow focusing of the image.
- The **fine-adjustment knob** moves the body tube slightly to bring the image into sharper focus.
- The **stage** supports a slide.
- **Stage clips** hold the slide in place for viewing.
- The **diaphragm** controls the amount of light coming through the stage.
- The light source provides a **light** for viewing the slide.
- The **arm** supports the body tube.
- The **base** supports the microscope.

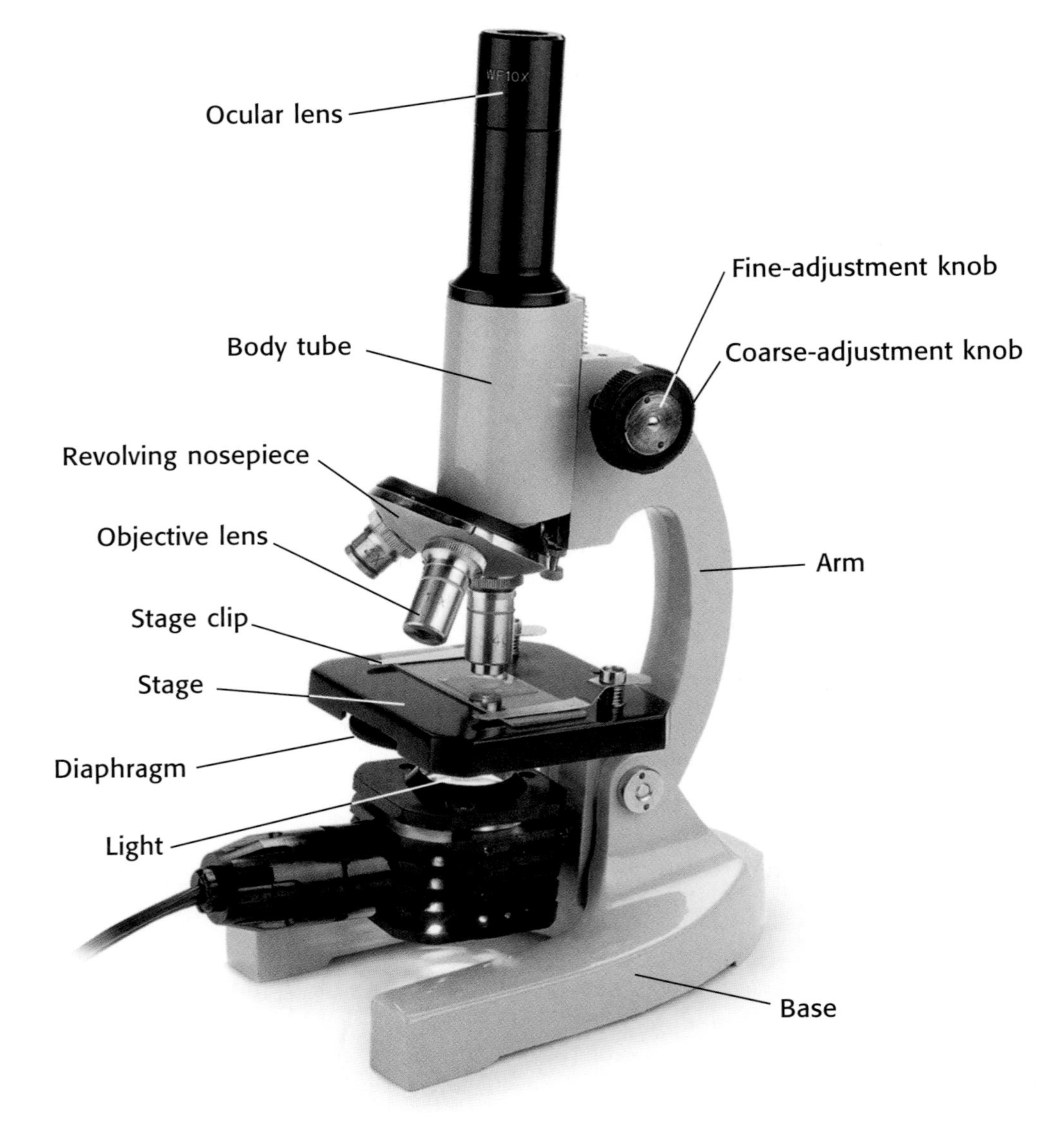

Proper Use of the Compound Light Microscope

1. Carry the microscope to your lab table using both hands. Place one hand beneath the base, and use the other hand to hold the arm of the microscope. Hold the microscope close to your body while moving it to your lab table.
2. Place the microscope on the lab table at least 5 cm from the edge of the table.
3. Check to see what type of light source is used by your microscope. If the microscope has a lamp, plug it in, making sure that the cord is out of the way. If the microscope has a mirror, adjust it to reflect light through the hole in the stage.
   Caution: If your microscope has a mirror, do not use direct sunlight as a light source. Direct sunlight can damage your eyes.
4. Always begin work with the low-power objective in line with the body tube. Adjust the revolving nosepiece.
5. Place a prepared slide over the hole in the stage. Secure the slide with the stage clips.
6. Look through the ocular lens. Move the diaphragm to adjust the amount of light coming through the stage.
7. Look at the stage from eye level. Slowly turn the coarse adjustment to lower the objective until it almost touches the slide. Do not allow the objective to touch the slide.
8. Look through the ocular lens. Turn the coarse adjustment to raise the low-power objective until the image is in focus. Always focus by raising the objective away from the slide. *Never focus the objective downward.* Use the fine adjustment to sharpen the focus. Keep both eyes open while viewing a slide.
9. Make sure that the image is exactly in the center of your field of vision. Then switch to the high-power objective. Focus the image, using only the fine adjustment. *Never use the coarse adjustment at high power.*
10. When you are finished using the microscope, remove the slide. Clean the ocular lens and objective lenses with lens paper. Return the microscope to its storage area. Remember, you should use both hands to carry the microscope.

Making a Wet Mount

1. Use lens paper to clean a glass slide and a coverslip.
2. Place the specimen you wish to observe in the center of the slide.
3. Using a medicine dropper, place one drop of water on the specimen.
4. Hold the coverslip at the edge of the water and at a 45° angle to the slide. Make sure that the water runs along the edge of the coverslip.
5. Lower the coverslip slowly to avoid trapping air bubbles.
6. Water might evaporate from the slide as you work. Add more water to keep the specimen fresh. Place the tip of the medicine dropper next to the edge of the coverslip. Add a drop of water. (You can also use this method to add stain or solutions to a wet mount.) Remove excess water from the slide by using the corner of a paper towel as a blotter. Do not lift the coverslip to add or remove water.

Glossary

A

abiotic describes nonliving factors in the environment (4, 48)

acid precipitation precipitation that contains acids due to air pollution (106)

adaptation a characteristic that helps an organism survive in its environment (16)

algae (AL JEE) protists that convert the sun's energy into food through photosynthesis (6, 17)

alien an organism that makes a home for itself in a new place (78)

Animalia the classification kingdom containing complex, multicellular organisms that lack cell walls, are usually able to move around, and possess nervous systems that help them be aware of and react to their surroundings (157)

atmosphere a mixture of gases that surrounds a planet, such as Earth (74, 76)

atom the smallest part of an element that has all of the properties of that element (154)

B

biodegradable capable of being broken down by the environment (80)

biodiversity the number and variety of living things (79)

biome a large region characterized by a specific type of climate and certain types of plant and animal communities (48)

biosphere the part of the Earth where life exists (7)

biotic describes living factors in the environment (4)

C

carbon cycle the movement of carbon from the nonliving environment into living things and then back into the nonliving environment (31)

carnivore a consumer that eats animals (9)

carrying capacity the largest population that a given environment can support over a long period of time (15)

coal a solid fossil fuel formed underground from buried, decomposed plant material (102)

coevolution (KOH ev uh LOO shuhn) the long-term changes that take place in two species because of their close interactions with one another (18)

community all of the populations of different species that live and interact in an area (6)

competition two or more species or individuals trying to use the same limited resource (15)

compound light microscope a microscope that consists of a tube with lenses, a stage, and a light source (158)

conifer a tree that produces seeds in cones (50)

conservation the wise use of and preservation of natural resources (81)

consumer an organism that eats producers or other organisms for energy (9)

D

deciduous describes trees with leaves that change color in autumn and fall off in winter (49)

decomposer an organism that gets energy by breaking down the remains of dead organisms or animal wastes and consuming or absorbing the nutrients (9)

decomposition the breakdown of dead materials into carbon dioxide and water (32)

deep-water zone the zone of a lake or pond below the open-water zone where no light reaches (61)

deforestation the clearing of forest lands (79)

desert a hot, dry biome inhabited by organisms adapted to survive high daytime temperatures and long periods without rain (53)

diversity a measure of the number of species an area contains (51)

E

energy pyramid a diagram shaped like a triangle that shows the loss of energy at each level of the food chain (11)

energy resource a natural resource that humans use to produce energy (101)

estuary an area where fresh water from streams and rivers spills into the ocean (59)

evaporation the change of state from liquid to vapor (30)

extinct describes a species of organism that has died out completely (85)

F

food chain a diagram that represents how the energy in food molecules flows from one organism to the next (10)

food web a complex diagram representing the many energy pathways in a real ecosystem (10)

fossil fuel a nonrenewable energy resource that forms in the Earth's crust over millions of years from the buried remains of once-living organisms (101)

G

gasohol a mixture of gasoline and alcohol that is burned as a fuel (114)

geothermal energy energy from within the Earth (115)

global warming a rise in average global temperatures (31, 76)

ground water water stored in underground caverns or porous rock (31)

H

habitat the environment where an organism lives (12)

herbivore a consumer that eats plants (9)

host an organism on which a parasite lives (18)

hydroelectric energy electricity produced by falling water (113)

hypothesis a possible explanation or answer to a question (142–144)

K

kingdom the most general of the seven levels of classification (156–157)

L

limiting factor a needed resource that is in limited supply (14)

littoral zone the zone of a lake or pond closest to the edge of the land (61)

M

marine describes an ecosystem based on salty water (55)

marsh a treeless wetland ecosystem where such plants as cattails and rushes grow (62)

matter anything that occupies space and has mass (30)

meter the basic unit of length in the SI system (139)

mutualism (MYOO choo uhl IZ uhm) a symbiotic relationship in which both organisms benefit (17)

N

natural gas a gaseous fossil fuel (102)

natural resource any natural substance, organism, or energy form that living things use (98)

niche an organism's way of life and its relationships with its abiotic and biotic environments (12)

nitrogen cycle the movement of nitrogen from the nonliving environment into living organisms and back again (32)

nitrogen fixation the process of changing nitrogen gas into forms that plants can use (33)

nonrenewable resource a natural resource that cannot be replaced or that can be replaced only over thousands or millions of years (77, 82, 99-100)

nuclear energy the form of energy associated with changes in the nucleus of an atom; an alternative energy resource (108)

nuclear fusion the process by which two or more nuclei with small masses join together, or fuse, to form a larger, more massive nucleus, along with the production of energy (109)

O

observation any use of the senses to gather information (144)

omnivore a consumer that eats a variety of organisms (9)

open-water zone the zone of a lake or pond that extends from the littoral zone out across the top of the water and that is only as deep as light can reach through the water (61)

overpopulation a condition that occurs when the number of individuals within an environment becomes so large that there are not enough resources to support them all (78)

ozone a gas molecule that is made up of three oxygen atoms; absorbs ultraviolet radiation from the sun (75–76)

P

parasite an organism that feeds on another living creature, usually without killing it (18)

parasitism (PAR uh SIET IZ uhm) a symbiotic association in which one organism benefits while the other is harmed (18)

permafrost the permanently frozen ground below the soil surface in the arctic tundra (54)

petroleum an oily mixture of flammable organic compounds from which liquid fossil fuels and other products are separated; crude oil (101)

phytoplankton (FITE oh PLANK tuhn) a microscopic photosynthetic organism that floats near the surface of the ocean (55)

pioneer species the first organisms to grow in an area undergoing ecological succession; usually lichens in primary succession and fast-growing, weedy plants in secondary succession (35)

plankton very small organisms floating at or near the ocean's surface that form the base of the ocean's food web (55)

Plantae the kingdom that contains plants—complex, multicellular organisms that are usually green and use the sun's energy to make sugar by photosynthesis (157)

pollutant a harmful substance in an environment (74)

pollution the presence of harmful substances in an environment (74)

population a group of individuals of the same species that live together in the same area at the same time (6)

precipitation water that moves from the atmosphere to the land and ocean, including rain, snow, sleet, and hail (30)

predator an organism that eats other organisms (16)

prey an organism that is eaten by another organism (16)

producer organisms that make their own food, usually by using the energy from sunlight to make sugar (8)

Protista a kingdom of eukaryotic single-celled or simple, multicellular organisms; kingdom Protista contains all eukaryotes that are not plants, animals, or fungi (156)

R

radiation energy transferred as waves on particles (108)

recycling the process of making new products from reprocessed used products (83, 100)

renewable resource a natural resource that can be used and replaced over a relatively short time period (77, 99)

resource recovery the process of transforming into usable products things normally thrown away (84)

S

savanna a tropical grassland biome with scattered clumps of trees (52)

scavenger an animal that feeds on the bodies of dead animals (9)

sediment fine particles of sand, dust, or mud that are deposited over time by wind or water (103)

smog a photochemical fog produced by the reaction of sunlight and air pollutants (107)

soil a loose mixture of small mineral fragments and organic material (35)

solar energy energy from the sun (109)

strip mining a process in which rock and soil are stripped from the Earth's surface to expose the underlying materials to be mined (105)

succession the gradual regrowth or development of a community of organisms over time (34)

swamp a wetland ecosystem in which trees and vines grow (63)

symbiosis (SIM bie OH sis) a close, long-term association between two or more species (17)

T

temperature a measure of how hot or cold something is (140)

toxic poisonous (75)

tributary a small stream or river that flows into a larger one (60)

tundra a far-northern biome characterized by long, cold winters, permafrost, and few trees (54)

W

water cycle the movement of water between the ocean, atmosphere, land, and living things (30)

wetland an area of land where the water level is near or above the surface of the ground for most of the year (62)

wind energy energy in wind (112)

Z

zooplankton (ZOH oh PLANGK tuhn) protozoa that, along with the phytoplankton they consume, form the base of the ocean's food web (55)

Index

Boldface numbers refer to an illustration on that page.

A

B

C

D

E

N

O

P

R

S

INDEX

T

U

V

W

X

Z

Credits

Abbreviations used: (t) top, (c) center, (b) bottom, (l) left, (r) right, (bkgd) background

ILLUSTRATIONS

All illustrations, unless otherwise noted below by Holt, Rinehart and Winston.

Table of Contents Page iv(cl) Will Nelson/Sweet Reps

Chapter One Page 4 (b); 5 (l), Will Nelson/Sweet Reps; 6-7 (b) John White/The Neis Group; 8-9 (b) Will Nelson/Sweet Reps; 10 (b), John White/The Neis Group; 11 (b), Will Nelson/Sweet Reps; 13 (br), Will Nelson/Sweet Reps; 14 (bl), Blake Thornton/Rita Marie; 19 (cr), Mike Wepplo/Das Group; 22 (cl), Will Nelson/Sweet Reps; 25 (cr), Jared Schneidman Design.

Chapter Two Page 30 (b), Robert Hynes/Mendola Artists; 31 (b), Robert Hynes/Mendola Artists; 32 (b), Robert Hynes/Mendola Artists; 35, Robert Hynes/Mendola Artists; 36, Robert Hynes/Mendola Artists; 40 (br), Robert Hynes/Mendola Artists; 42 (br), Robert Hynes/Mendola Artists; 45 (cl), MapQuest.com.

Chapter Three Page 48 (b), MapQuest.com; 49 (b), Will Nelson/Sweet Reps; 50 (b), Will Nelson/Sweet Reps; 51 (b), Will Nelson/Sweet Reps; 53 (b), Will Nelson/Sweet Reps; 56-57, Yuan Lee; 60 (bl), Will Nelson/Sweet Reps; 61 (br), Mark Heine; 64 (b), Carlyn Iverson/67 (cr), Mark Heine; 68 (bl), Will Nelson/Sweet Reps; 69 (tr,cr), Rob Schuster/Hankins and Tegenborg.

Chapter Four Page 75(tr), Peter Darro; 93 (t), John White/The Neis Group.

Chapter Five: Page 98(b), Uhl Studios, Inc.; 103(bl), Uhl Studios, Inc.; 104(l), Uhl Studios, Inc.; 105(t), MapQuest.com; 108(cl), Stephen Durke/Washington Artists; 111(tr), John Huxtable/Black Creative; 115(br), Uhl Studios, Inc.; 118(br), John Huxtable/Black Creative; 118(cr), Uhl Studios, Inc.; 121(cr), Sidney Jablonski.

LabBook Page 132 (r), John White/The Neis Group.

Appendix Page 140 (t), Terry Guyer; 144 (b), Mark Mille/Sharon Langley; 152-153, Kristy Sprott; 154 (bl), Stephen Durke/Washington Artists; 155 (tl,c) Stephen Durke/Washington Artists; 155 (b), Bruce Burdick; 155 (cl), Stephen Durke/Washington Artists.

PHOTOGRAPHY

Cover and Title page: Secret Sea Visions/Peter Arnold, Inc.

Feature Borders: Unless otherwise noted below, all images ©2001 PhotoDisc/HRW. "Across the Sciences" 70, all images by HRW; "Careers" 71, 94, sand bkgd and Saturn, Corbis; DNA, Morgan Cain & Associates; scuba gear, ©1997 Radlund & Associates for Artville; "Eureka" 123, ©2001 PhotoDisc/HRW; "Eye on the Environment" 27, 122, clouds and sea in bkgd, HRW; bkgd grass, red eyed frog, Corbis ; hawks, pelican, Animals Animals/Earth Scenes; rat, Visuals Unlimited/John Grelach; endangered flower, Dan Suzio/Photo Researchers, Inc.; "Health Watch" 26, dumbbell, Sam Dudgeon/HRW Photo; aloe vera, EKG, Victoria Smith/HRW Photo; basketball, ©1997 Radlund & Associates for Artville; shoes, bubbles, Greg Geisler; "Scientific Debate" 95, Sam Dudgeon/HRW Photo; "Weird Science" 44, mite, David Burder/Stone; atom balls, J/B Woolsey Associates; walking stick, turtle, EclectiCollection.

Table of Contents: iv(tl), Sanford D. Porter/U.S. Department of Agriculture; iv(cl), Jeff Hunter/Image Bank; iv(b), Sylvian Coffie/Stone; v(tr), Owen Garrett/Centre for Atmospheric Science at Cambridge University, UK/NASA; v(cr), J. Contreras Chacel/International Stock; v(b), Peter Van Steen/HRW Photo' vi(tl), SuperStock; vi(cl), Peter Van Steen/HRW Photo; vi(b), Peter Van Steen/HRW Photo; vii(tr), E. R. Degginger/Color-Pic, Inc.; vii(cr), Lincoln P. Brower; vii(br), Gay Bumgarner/Stone.

Chapter One: pp. 2-3 Hans Reinhard/Bruce Coleman, Inc.; 3 HRW Photo; 12(bl), Gamma-Liaison; 12(tl), Laguna Photo/Liaison International; 13(cr), Jeff Lepore/Photo Researchers, Inc.; 14(c), Jeff Foott/AUSCAPE; 15 Ross Hamilton/Stone; 16(tl), Visuals Unlimited/Gerald & Buff Corsi; 16(bl), Hans Pfletschinger/Peter Arnold; 17(b), Ed Robinson/Tom Stack & Associates; 17(cr), Telegraph Color Library/FPG; 17(tr), Peter Parks/Animals Animals Earth Scenes; 18(tl), Gay Bumgarner/Stone; 18(bl), Carol Hughes/Bruce Coleman; 19(tr), CSIRO Wildlife & Ecology; 21 Victoria Smith/HRW Photo; 23(cr), Gay Bumgarner/Stone; 26(tc), Darlyne Murawski/National Geographic Society Image Collection; 27(tr), Sanford D. Porter/U.S. Department of Agriculture

Chapter Two: pp 28-29 John D. Dawson/USPS; 29 HRW Photo; 33 Ray Pfortner/Peter Arnold, Inc.; 34(t), Diana L. Stratton/Tom Stack; 34(b), Stan Osolinski/FPG; 37 Kim Heacox/DRK Photo; 39(tr) Kenneth Gabrielson/Liaison International; 39(tl) Doug Sokell/Visuals Unlimited; 39(b) Larry Nielsen/Peter Arnold 40(tl), Ray Pfortner/Peter Arnold, Inc.; 41 Kim Heacox/DRK Photo; 44 Charles O'Rear/Westlight

Chapter Three: pp. 46-47 VCG/FPG International; 47 HRW Photo; 52(c), Grant Heilman; 52(b), Tom Brakefield/Bruce Coleman; 54 Kathy Bushue/Stone; 55(bc), Stuart Westmorland/Stone; 55(bl), Manfred Kage/Peter Arnold; 58(tl), Jeff Hunter/Image Bank; 58(bl), Zig Leszczynski/Animals Animals; 59(tr) Johnny Johnson/DRK Photo; 59(cr), Nancy Sefton/Photo Researchers, Inc.; 62(b), Dwight Kuhn; 62(tr), Unicorn Stock Photos; 63(cr), Don & Pat Valenti/DRK Photo; 63(tr), Hardie Truesdale; 67(tl), Jeff Hunter/Image Bank; 70 Dr. Verena Tunnicliffe; 71(tl, br), Lincoln P. Brower

Chapter Four: pp. 72-73 Peter Cade/Stone; 73 HRW Photo; 74(c), Grant Heilman; 74(bl), Arthur Tilley/Stone; 75(br), Ken Griffiths/Stone; 76(c), Roy Morsch/Stock Market; 76(tl), Owen Garrett/Centre for Atmospheric Science at Cambridge University, UK/NASA; 77 Jacques Jangoux/Stone; 78(tl), Runk/ Schoenberger/Grant Heilman; 78(cl), John Eastcott/VVA Momatiuk/Woodfin Camp; 79(tr), Rex Ziak/Stone; 79(br), Martin Rogers/Uniphoto; 80(cl), Fed Bavendam/Peter Arnold; 82(tl), Argonne National Laboratory; 82(bl), Emile Luider/Rapho/Liaison; 83(tr), PhotoEdit; 83(bl), Kay Park-Rec Co; 83(bc), J. Contreras Chacel/International Stock; 84(cl), Martin Bond/Science Photo Library/Photo Researchers, Inc.; 85(cl), Uniphoto; 85(br), K. W. Fink/Bruce Coleman; 85(tr), Toyohiro Yamada/FPG; 86 Stephen J. Krasemann/DRK Photo; 87(tr), Will & Deni McIntyre/Stone; 87(cr), Stephen J. Krasemann/DRK Photo; 90(cl), Arthur Tilley/Stone; 90(cr), K. W. Fink/Bruce Coleman, Inc.; 92 Runk/Schoenberger/ Grant Heilman; 94(tl, br), Karen M. Allen; 95(bc), Art Wolfe

Chapter Five: pp. 96-97 Novovitch/Liaison Agency; 96 Roger Ressmeyer/ Corbis; 97 HRW Photo; 98(c), Andy Christiansen/HRW Photo; 98(l), John Blaustein/Liaison Agency; 98(r), Mark Lewis/Stone; 99(tc), James Randklev/Stone; 99(tr), Bruce Hands/Stone; 99(br), John Zoiner Photographer; 99(bl), Ed Malles/Liaison Agency; 100(cl), Andy Christiansen/HRW Photo; 101 Telegraph Colour Library/FPG International; 102(cr, bl), John Zoiner; 104(t), Horst Schafer/Peter Arnold, Inc.; 104(ct), Paolo Koch/Photo Researchers, Inc.; 104(cb), Brian Parker/Tom Stack & Associates; 104(b), C. Kuhn/Image Bank; 105(bl), Mark A. Leman/Stone; 105(br), Tim Eagan/Woodfin Camp & Associates; 106(tr), Adam Hart-Davis/Science Photo Library/Photo Researchers, Inc.; 106(bl), James Stanfield/National Geographic Image Collection; 107(tc, tr), Victoria Smith/ HRW Photo; 108(bl), Sylvain Coffie/Stone; 109(tr), Tom Myers/ Photo Researchers, Inc.; 110(c), Alex Bartel/Science Photo Library/Photo Researchers, Inc.; 110(bl), Joyce Photographics/Photo Researchers,Inc.; 111(bl), Hank Morgan/Science Source/Photo Researchers, Inc.; 112, Mark Lewis/Liason International; 113(tr), Craig Sands/National Geographic Image Collection; 113(cl), Tom Bean; 114(cr), G.R. Roberts Photo Library; 116 Richard Hutchings/HRW Photo; 120(tl), John Blaustein/Liaison Agency; 120(tr), Tom Myers/Photo Researchers, Inc.; 122(t), SuperStock; 122(c), Bedford Recycled Plastic Timbers; 122(b), Kay Park-Rec Co; 123 Culver Pictures, Inc.

Labook: "LabBook Header": "L", Corbis, "a", Letraset-Phototone, "b" and "B", HRW, "o" and "k", Images ©2001 PhotoDisc/HRW, Inc. 125(cl), Michelle Bridwell/HRW Photo; 125(br), Image ©2001 Photodisc, Inc.; 126(bl), Stephanie Morris/HRW Photo; 127(tr), Jana Birchum/HRW Photo; 135(tr), Tom Bean/DRK Photo; 135(br), Darrell Gulin/DRK Photo, 136(br), Mark Heine.

Appendix: p. 158 CENCO

Sam Dudgeon/HRW Photos: all Systems of the Body background photos, p. viii-1, 80(tr), 87(br), 107(cr, c), 109(b), 114(tl), 117, 124, 125(bc), 126(br, tl, tr), 127(tl), 128(bc, tr), 129(tr, br), 130, 131(tr, br), 137, 141(br)

Peter Van Steen/HRW Photos: p. 38, 81(bl, bc, br), 84(tl), 89, 91, 127(b), 133, 134, 141(tr)

John Langford/HRW Photos: p. 125(tr)

Acknowledgements continued from page iii.

Alyson Mike
Science Teacher
East Valley Middle School
East Helena, Montana

Donna Norwood
Science Teacher and Dept. Chair
Monroe Middle School
Charlotte, North Carolina

James B. Pulley
Former Science Teacher
Liberty High School
Liberty, Missouri

Terry J. Rakes
Science Teacher
Elmwood Junior High School
Rogers, Arkansas

Elizabeth Rustad
Science Teacher
Crane Middle School
Yuma, Arizona

Debra A. Sampson
Science Teacher
Booker T. Washington Middle School
Elgin, Texas

Charles Schindler
Curriculum Advisor
San Bernadino City Unified Schools
San Bernadino, California

Bert J. Sherwood
Science Teacher
Socorro Middle School
El Paso, Texas

Patricia McFarlane Soto
Science Teacher and Dept. Chair
G. W. Carver Middle School
Miami, Florida

David M. Sparks
Science Teacher
Redwater Junior High School
Redwater, Texas

Elizabeth Truax
Science Teacher
Lewiston-Porter Central School
Lewiston, New York

Ivora Washington
Science Teacher and Dept. Chair
Hyattsville Middle School
Washington, D.C.

Elsie N. Waynes
Science Teacher and Dept. Chair
R. H. Terrell Junior High School
Washington, D.C.

Nancy Wesorick
Science and Math Teacher
Sunset Middle School
Longmont, Colorado

Alexis S. Wright
Middle School Science Coordinator
Rye Country Day School
Rye, New York

John Zambo
Science Teacher
E. Ustach Middle School
Modesto, California

Gordon Zibelman
Science Teacher
Drexel Hill Middle School
Drexell Hill, Pennsylvania

SELF-CHECK

Self-Check Answers

Chapter 1—Interactions of Living Things

Page 9: Humans are omnivores. An omnivore eats both plants and animals. Humans can eat meat and vegetables as well as animal products, such as milk and eggs, and plant products such as grains and fruit.

Page 10: A food chain shows how energy moves in one direction from one organism to the next. A food web shows that there are many energy pathways between organisms.

Page 15: 1. If an area has enough water to support 10 organisms, any additional organisms will cause some to go without water and move away, or die. 2. Weather favorable for growing the food that deer eat will allow the forest to support more deer.

Chapter 2—Cycles in Nature

Page 36: The main difference between primary and secondary succession is that primary succession begins with the formation of soil. Secondary succession begins on preexisting soil, such as when an existing community is disrupted by a natural disaster or by farming. Pioneer species in primary succession are usually lichens, which begin the formation of soil. Pioneer species in secondary succession are usually seed plants, which germinate and take root in the soil.

Chapter 3—The Earth's Ecosystems

Page 53: Deciduous forests tend to exist in mid-latitude or temperate regions, while coniferous forests tend to exist in higher, colder latitudes, closer to the poles.

Page 58: Answers include: the amount of sunlight penetrating the water, its distance from land, the depth of the water, the salinity of the water, and the water's temperature. 2. There are several possible answers. Some organisms are adapted for catching prey at great depths; some feed on dead plankton and larger organisms that filter down from above; and some, such as the bacteria around thermal vents, make food from chemicals in the water.

Chapter 4—Environmental Problems and Solutions

Page 77: 1. We use nonrenewable resources when we burn fossil fuels when driving or riding in a car or burning coal for heat. When we use minerals that are mined, we are using a nonrenewable resource. Pumping ground water is another use of a nonrenewable resource, if the water is used faster than it is replenished. 2. If a nonrenewable resource is used up, we can no longer rely on that resource. Certain oil and coal deposits have been building since life began on the planet. It may take hundreds of years to replace a mature forest that can be cut in a day.

Page 83: 1. Turn off lights, CD players, radios, and computers when leaving a room. Set thermostats a little lower in the winter (wear sweaters). Don't leave the refrigerator door open while deciding what you want. 2. plastic bags, rechargeable batteries, water, clothing, toys; The difference between a reused and a recycled object is that a reused article may be cleaned but is basically unchanged. A recycled article has been broken down and re-formed into another usable product.

Chapter 5—Energy Resources

Page 113: Both devices harness energy from falling water.